MOTOR VEHICLE STUDIES

Terry Hewitt

£7.90

CW01574665

UNWIN HYMAN

Published by
UNWIN HYMAN LIMITED
15/17 Broadwick Street
London W1V 1FP

British Library Cataloguing in Publication Data

Hewitt, Terry
Motor Vehicle Studies.
1. Motor Vehicles. Engines
1. Title
629.2'5
ISBN 0 7135 2846 X

Design by Geoffrey Wadsley
Artwork by Jerry Collins
Cartoons by Shaun Williams
Typeset by Cambridge Photosetting Services
Printed and bound by The Alden Press, Oxford

Contents

Preface

This book attempts to make Motor Vehicle Studies accessible to a wide range of students. It is hoped that students of GCSE, C and G, TEC and RTIB courses particularly will find it useful. It will also prove relevant to a variety of modular courses in schools and colleges.

No single book can cover all aspects of the subject, and it is intended that work in this book should be supplemented by student reference to other texts, handbooks, manuals and associated resources. The language used is straightforward and the diagram as simple as possible without sacrificing technical accuracy. Practical experience of handling tools and components is essential to understanding, and reference is made to appropriate practical activities.

The page-per-topic layout will make the book easy to use, the questions should provoke some thought, and the end-of-chapter questions will stimulate revision of work that has gone before.

T.H.

Acknowledgements

I would like to thank many people for their enthusiasm and assistance. Particular thanks are due to Dick Bateman for his invaluable expertise and encouragement; Clive Warren for his technical advice; Mike Jones for his confidence and support; the students of Sir Bernard Lovell School for their constructive comments. On the publishing front, Chris Baker, Chris Blake, Tess Gonet and June Thompson have all helped enormously. Finally, a special thank you to my family – Sue, Ben, Jess and Lucy.

Photo credits

Aerofilms: Fig 48(g)
Austin Rover: p. 87
BBC Hulton Picture Library: Figs 34, 39(b)(i), 42(c), 48(d)
Deborah Bulbeck: Figs 40, 45, 48(a), 48(b)
Dunlop Tyres: Fig. 26(c)
EMAP National Publications Limited: Figs 43(a), 43(c), 43(e), 43(f), 43(h), 43(l)
Esso Petroleum Company Limited: p. 41
Fay Godwin: p. 1
Ford Motor Company Ltd: Figs 47(a), 47(b), pp. 53, 92 and 115
Gary Weaser: p. 5
Haymarket Motoring Photo Library: pp. 17, 29, Figs 26(b), 26(c), 39(a)(i), 39(a)(ii), 39(b)(ii), 41(b), 42(d), 42(e), 43.2, 43(b), 43(d), 46, p. 79
Hewlett-Packard: Fig. 39(c), p. 87
Hitachi Electronics: p. 71
Honda UK Limited: Figs 43(i), 43(j), 43(k), 43(m)
Leyland DAF Limited: Fig. 24
Motorola, Automotive Industrial & Electronics Group: p. 71
National Express Coaches: Fig. 38(e)
National Motor Museum: Figs 38(a), 38(d), 38(g), 38(h), 41(d), 41(e), 42(a), 42(b), 42(f), 42(i), 42(k), 48(e)
Pirelli tyres: p. 61
Quadrant Picture Library: Figs 38(b), 38(c), 38(f), 41(a), 41(c), 42(g), 42(h), 42(j), 48(c), 48(f), 48(h), 48(i), p. 101
Steve Ashton: Figs 43(g), 43(n), 43(o), 43(p), 43(q), 43(r)

Front cover:
London Women's Motor Mechanic Project
Spectrum

Safety

A vital aspect of any workplace which is largely a matter of commonsense and an awareness of potential dangers to self and others.

1. Safety in the workshop

Workshops are potentially very dangerous places. Safety is important. Commonsense rules and a little thought will make a workshop safe. You need to be aware of dangers to yourself and to others and have correct

- Clothing,
- Behaviour,
- Method of working,
- Equipment.

Clothing

- Overalls keep your ordinary clothes clean and keep cuffs, ties, belts etc. from causing problems.
- Soft shoes are no protection against falling engines or gearboxes.
- Eye protection is vital for some jobs.
- Watches, bracelets and necklaces are obvious hazards. Can you think of any more?

Behaviour

- Fooling around in a workshop is stupid, selfish and senseless.
- *Never* run. Do not play practical jokes. Do not show off.
- Be aware of others working near you. Be safety conscious.
- *Always* report any accidents that do occur.

Method of working

- Use a barrier cream to protect hands from infection.
- Never rush a job. Ask if you are in any doubt about how to tackle a particular job. Use a manual or handbook.
- Be careful with: electrics, rotating parts, sharp edges, heavy components, fumes from battery or petrol, etc.
- Work methodically. Clear up when you have finished.

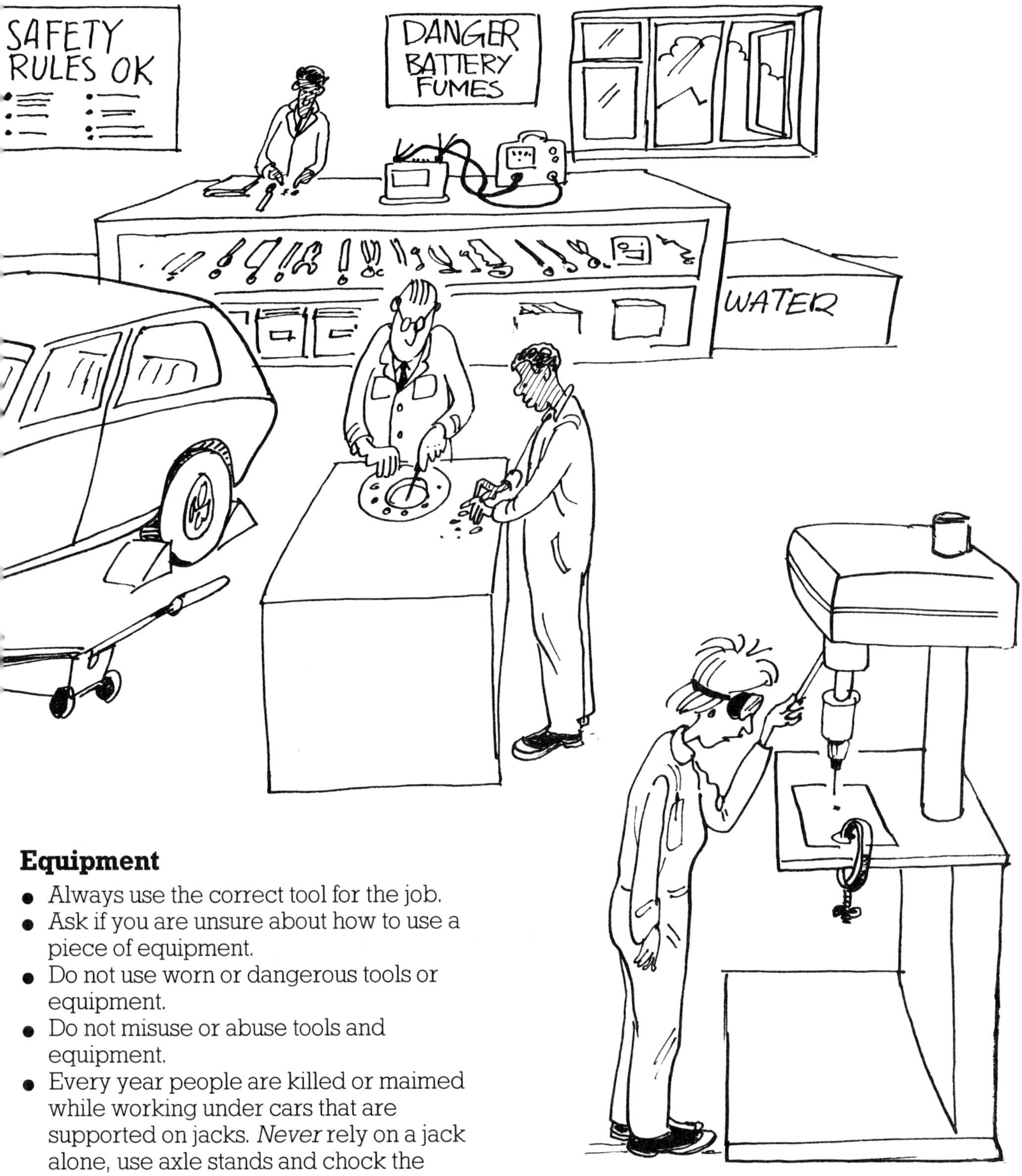

Equipment

- Always use the correct tool for the job.
- Ask if you are unsure about how to use a piece of equipment.
- Do not use worn or dangerous tools or equipment.
- Do not misuse or abuse tools and equipment.
- Every year people are killed or maimed while working under cars that are supported on jacks. *Never* rely on a jack alone, use axle stands and chock the wheels to make sure that the car will not move.

QUESTIONS

1 Make a list of *ten* safety rules for display in a workshop.

2 Write a short essay on the importance of safety in the workshop.

EXAMINATION QUESTIONS

WORKSHOP SAFETY

1 Name two types of fire extinguisher.

2 What precautions should be taken when working underneath a car?

3 Name *four* potentially dangerous substances which might be encountered when servicing or repairing a motor vehicle (*NEA*)

4 State two safety precautions that should be taken when working on a running engine. (*LEA*)

5 Explain why overalls and protective footwear should be worn at all times.

6 Why is barrier cream used when working in a workshop?

7 Before work is started on an engine the vehicle battery should be disconnected to reduce the risk of:
a) injury to the mechanic;
b) damage to the battery;
c) electric shocks;
d) the engine starting. (*NEA*)

8 What precautions should be taken when:
a) welding
b) drilling
c) carrying awkward heavy loads
d) carrying a battery
e) removing an engine from a car?

9 Explain, with examples, why the correct use of the correct tools and equipment for the job is important.

The engine

A device which converts the chemical energy stored in a fuel into heat energy inside a cylinder and converts this energy into mechanical energy through a piston, connecting rod and crankshaft.

2. Internal combustion

Internal and external combustion

Most motor vehicles use what is called an *internal combustion engine*. What does that mean? Well, let's start from the engine end. An *engine* changes energy into movement. In a car the energy is stored in the petrol. When the petrol burns in the engine the car can move. *Combustion* is another word for 'burning'. *Internal* means 'inside'. Internal combustion means 'burning inside'. An internal combustion engine works by burning petrol inside it.

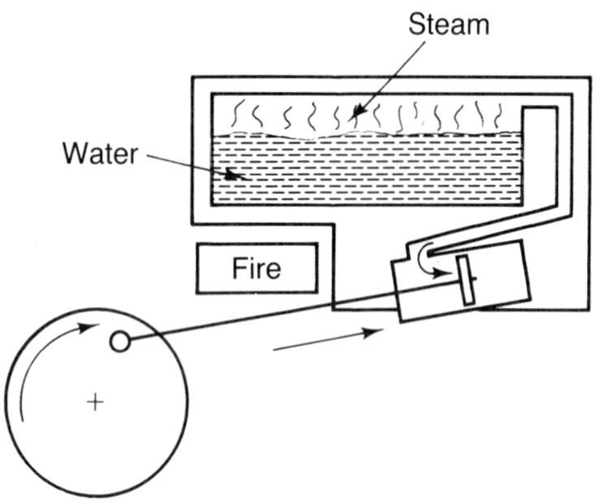

The first engines used in cars were *external* combustion engines. Coal was burned under a cylinder of water. The water boiled and turned to steam. The power of the steam was used to turn the wheels. [The person who shovelled the coal onto the fire was called a *chauvre feu* in France. This is how we get the word 'chauffeur'.] The pictures on the right show the difference between internal combustion and external combustion.

The technical explanation is that chemical energy (which is stored in the fuel) is changed to heat energy when the fuel combines with oxygen. The power released by this is increased by compressing the fuel and burning it in a gas tight cylinder.

Compression

Most internal combustion engines use petrol as a fuel.

If you pour petrol onto the floor and set light to it, it will burn. It will burn with a bright flame but no power. If you want power you must squash it into a small space (compress it). [Gunpowder is the same. On the ground it burns in a flash. Packed into a cardboard cylinder it explodes and bursts the cylinder.]

A simple internal combustion engine works like this:

1 Put petrol and air into a cylinder
2 Compress it
3 Burn it

Combustion needs air

Petrol will not burn without air. In a car a carburettor mixes petrol with just the right amount of air to make a fine spray of petrol/air mixture that is just right for burning.

If you stop air from getting to a fire it will go out.

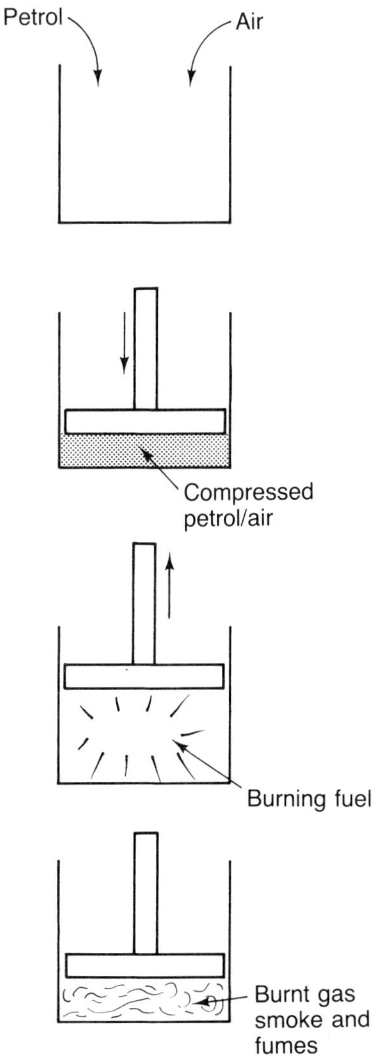

Fire extinguishers use foam or carbon dioxide (a gas which will not burn) to smother the fire. Garages and workshops often have fire buckets containing sand.

QUESTIONS

1 Explain the difference between internal combustion and external combustion.

2 Write four or five sentences about combustion. Mention petrol, air, carburettors and extinguishers.

WARNING
Petrol is *extremely* dangerous. Do not become one of the many people killed or injured by stupid 'experiments'!

3. The four-stroke cycle

Induction

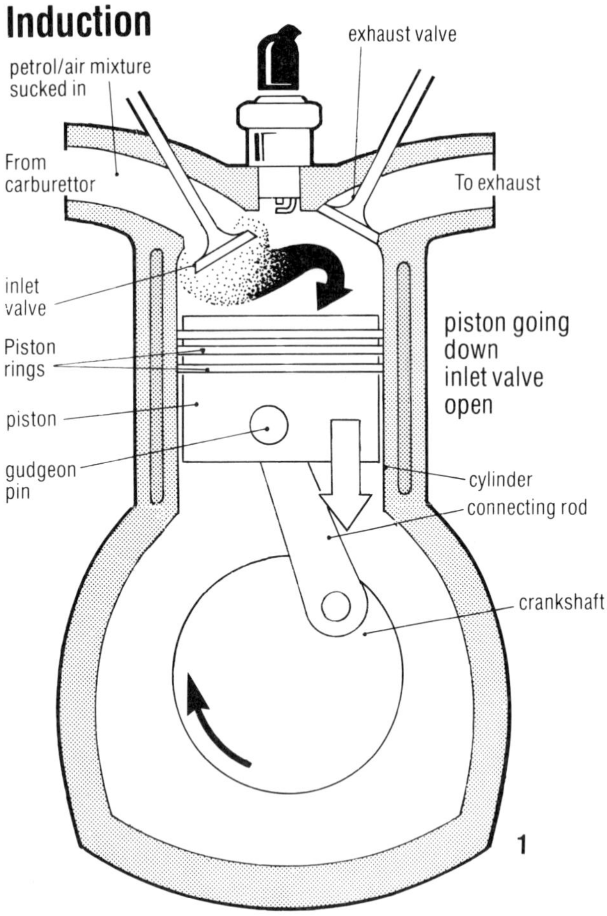

Compression

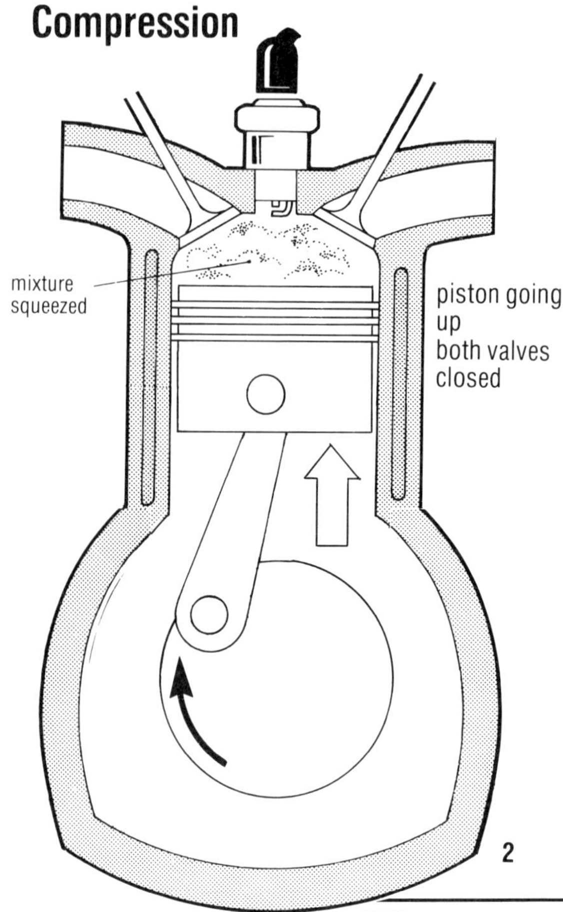

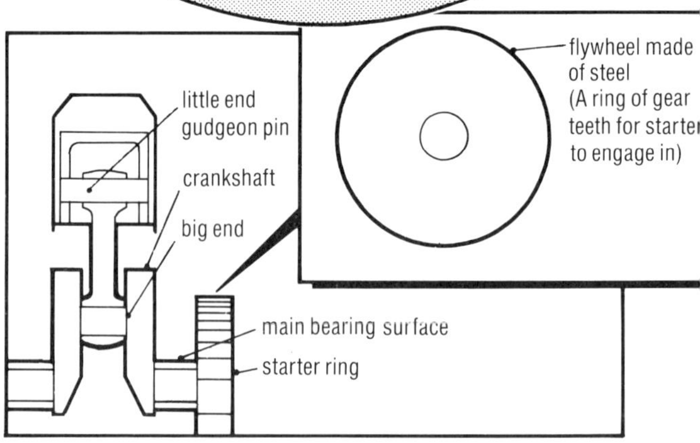

Most internal combustion engines work on what is called the four-stroke cycle. Combustion takes place in a steel or aluminium cylinder with an aluminium piston inside it. This piston can slide up and down. For one complete working cycle the piston moves four strokes (two up and two down).

- First a mixture of petrol and air is drawn into the cylinder.
- This is then compressed so that it burns with more power.
- Next the mixture is ignited.
- Finally, the smoke and fumes have to be pushed out before the cycle starts again.

How the four-stroke engine works

1 INDUCTION (suck)
As the piston moves down in the cylinder a valve opens and the piston draws a mixture of petrol and air into the cylinder. Towards the end of this stroke the inlet valve closes.

2 COMPRESSION (squeeze)
The piston moves up and, because both valves are closed, the petrol/air mixture is squeezed into the very small space between the top of the piston and the top of the cylinder.

3 POWER (bang)
When the piston is almost at the top of the compression stroke the spark plug ignites the mixture, which burns very rapidly and expands. This forces the piston down and gives the power to turn the crankshaft.

4 EXHAUST (blow)
Towards the end of the power stroke the exhaust valve opens and the piston moves up again. It pushes out the exhaust gases which are left after the mixture has burnt.

Power

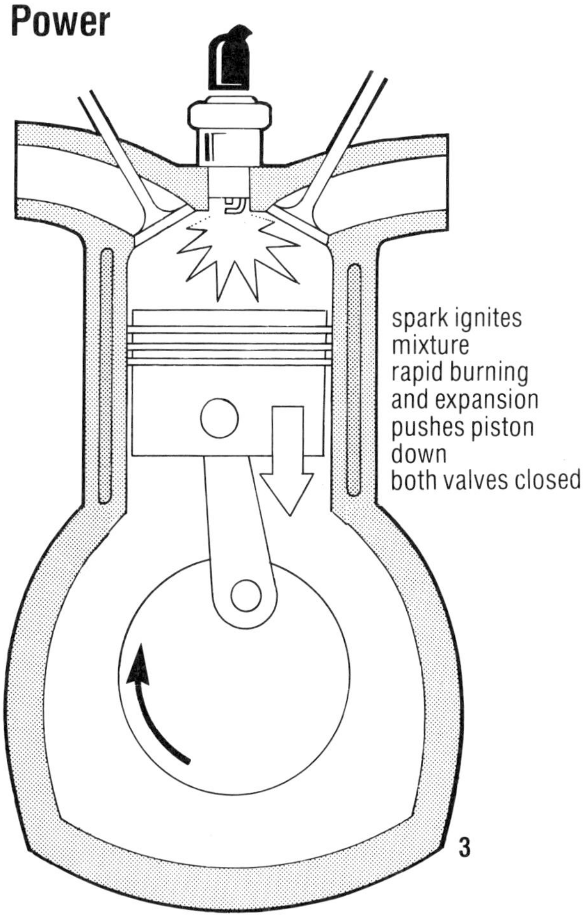

3

Exhaust

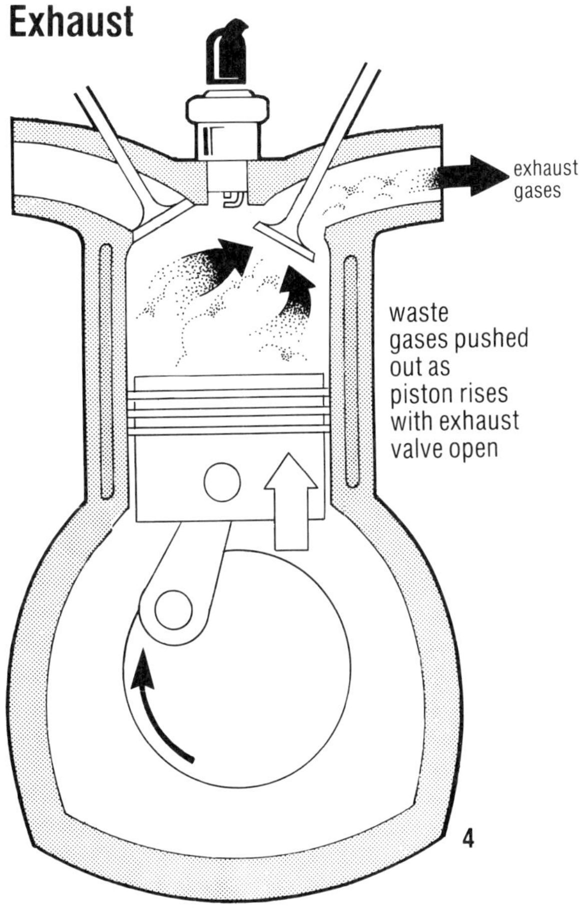

4

If you look carefully at diagrams 1 to 4 you will see that only one stroke actually produces power (no. 3). To keep the engine turning through the other strokes a heavy metal flywheel is fitted to the end of the crankshaft. Once this is turning it tends to keep turning and to smooth out the jerky 'one power stroke every four strokes' movement of the engine.

Remember that the diagrams show what happens inside one cylinder. A single cylinder four-stroke will produce enough power for a lawn mower, kart or motor cycle. Most cars have several cylinders (usually 4 or 6 but sometimes 8, or even 12). In these 'multi-cylinder' engines the power stroke happens at different times in each cylinder. The cycle also happens very quickly. If an engine is running at 6000 r.p.m. the piston goes up and down 100 times every second!

As the engine has to produce enough power to be able to pull a car which weighs about 1 tonne, four passengers, the family dog, a loaded roofrack and caravan along motorways and up hills the energy which is stored in petrol must be quite considerable!

QUESTIONS

1 Make neat diagrams of each of the four strokes of the four-stroke cycle.

2 Make a neat, labelled diagram of a section through an engine to show inlet valve, exhaust valve, spark plug, cylinder, piston, connecting rod, gudgeon pin and crankshaft.

Practical: Set valve clearance on one inlet and one exhaust valve.

4. The main parts of the engine

Cylinder block

Four-cylinder in-line block made of:
cast iron – strong and cheap
or
aluminium alloy – light but more expensive.

Piston (aluminium alloy)

Moves up and down in cylinder compressing mixture, being forced down by expanding gases and forcing out exhaust gases.

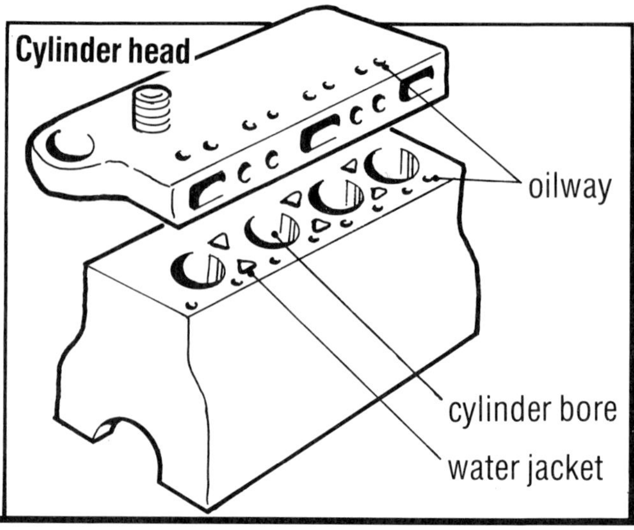

How to draw the block

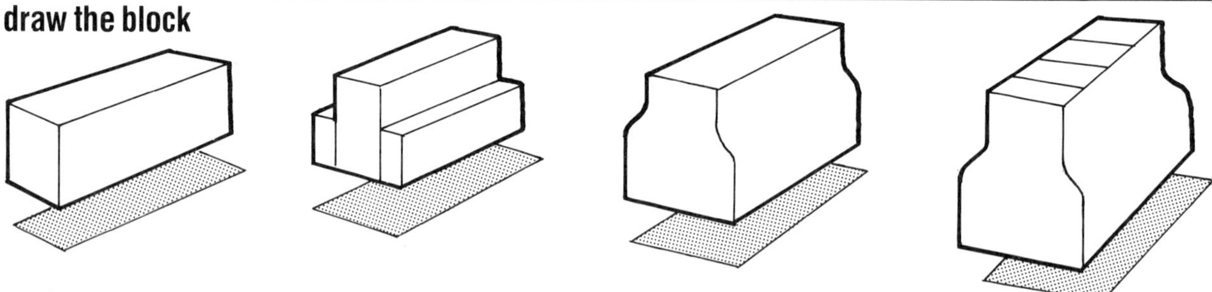

Connecting rod (forged steel)

Joins piston to crankshaft. It has two bearings: little end and big end.

Gudgeon pin (case hardened steel)

Connects piston to connecting rod.

Crankshaft (forged steel or cast iron)

Passes power of engine to flywheel. Crankshafts with five main bearings are more expensive than those with three.

Piston, connecting rod and crankshaft assembly

Changes reciprocating (up and down) motion of piston to rotary motion of crankshaft.

Camshaft (case hardened steel)

Controls the opening and closing of the valves. Driven by the crankshaft, one turn of camshaft to two turns of crankshaft. A sprocket on the end of the crankshaft drives a sprocket on the end of the camshaft by a timing chain. This makes sure that the valves open and close when the piston is in the right position.

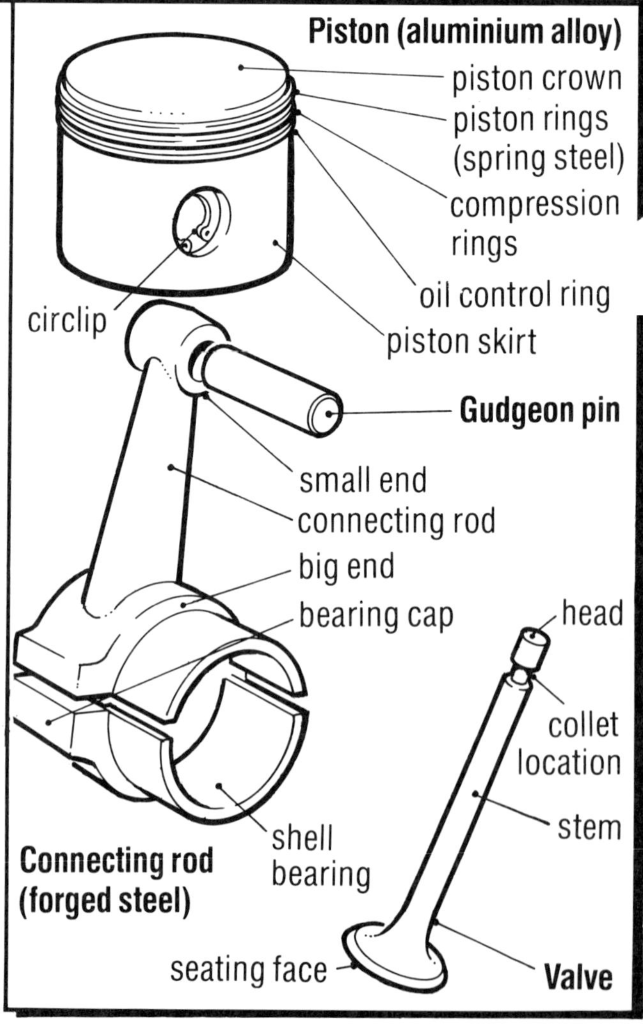

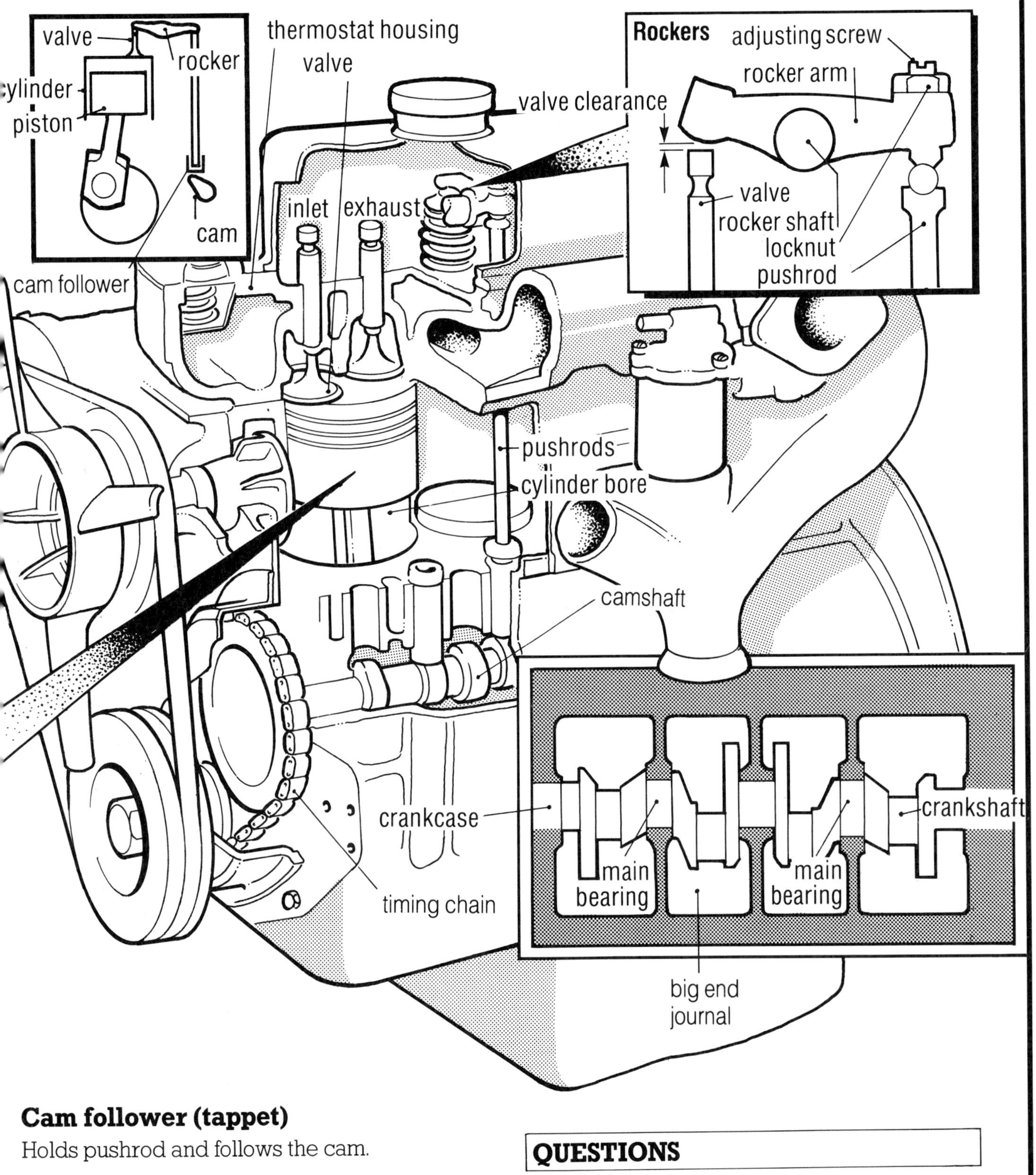

Cam follower (tappet)
Holds pushrod and follows the cam.

Rockers
Transmit movement of pushrod to the valves.

Valves
Allow fuel and air into cylinder and exhaust gases out; form an airtight seal when closed.

QUESTIONS

1 Copy the diagrams of the main parts of the engine.
2 Add short notes to explain the purpose of each component.

Practical: Remove cylinder head, remove, handlap and replace one valve, refit cylinder head.

5. Two-stroke engines

Four-stroke engines have only one power stroke every two turns of the crankshaft. Two-stroke engines use both sides of the piston and give a power stroke every single turn of the crankshaft.

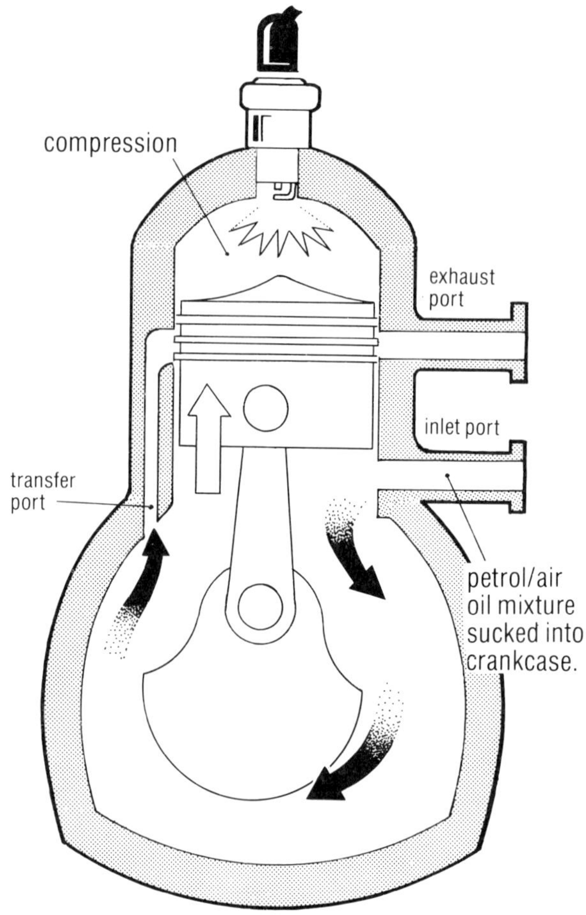

The two-stroke cycle

UP STROKE (induction and compression)

- *Below piston:* The inlet port is uncovered and a mixture of petrol, oil and air is sucked into the crankcase.
- *Above piston:* The mixture is compressed and the spark plug ignites it.

DOWN STROKE (power, transfer and exhaust)

- *Below piston:* Mixture is partly compressed and pushed through a transfer port into the cylinder above the piston.
- *Above piston:* Burning gases expand and push piston down. Exhaust gases are pushed out through the exhaust port, helped by the fresh mixture which is being pushed in through the transfer port.

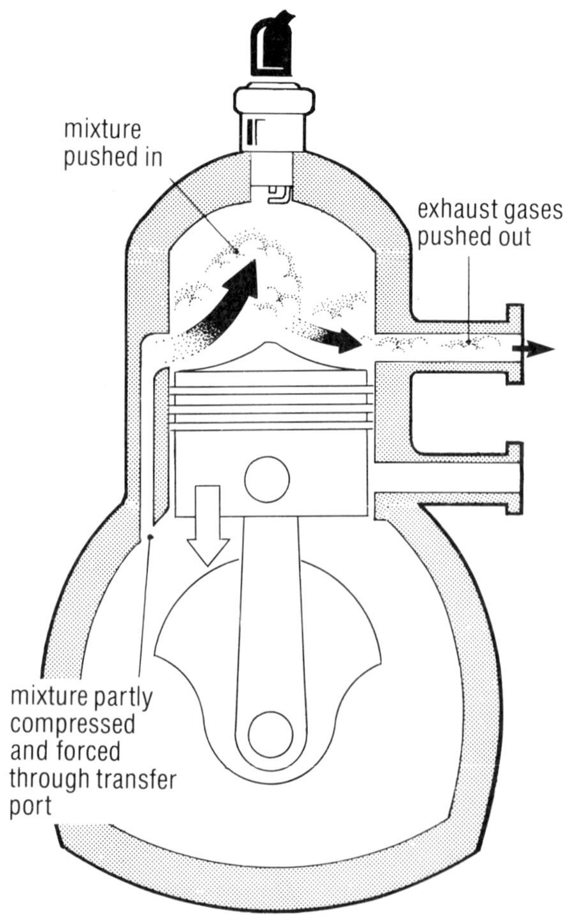

Advantages and disadvantages

Notice that two-stroke engines do not use valves. They have holes in the side of the cylinder wall called *ports.* As the piston moves these ports are uncovered and inlet mixture flows in or exhaust gas is forced out. Because there are no valves, camshafts, pushrods etc. two-stroke engines are simpler, lighter and cheaper than four-strokes.

Two-stroke engines use a fuel which is a mixture of petrol, oil and air. As the mixture goes through the crankcase the fine droplets of oil stick to the metal surfaces, lubricating them. This saves the cost and weight of oil pumps and filters. There is no valve gear to be lubricated; all of the moving parts are below the piston.

The two-stroke mixture of petrol, air and oil will not burn as well as petrol and air with no oil in it, so two-stroke engines are less efficient than four-stroke engines. Because the mixture has to pass through the crankcase, two strokes are better at lower speeds than at high speeds.

As the mixture enters the cylinder from the

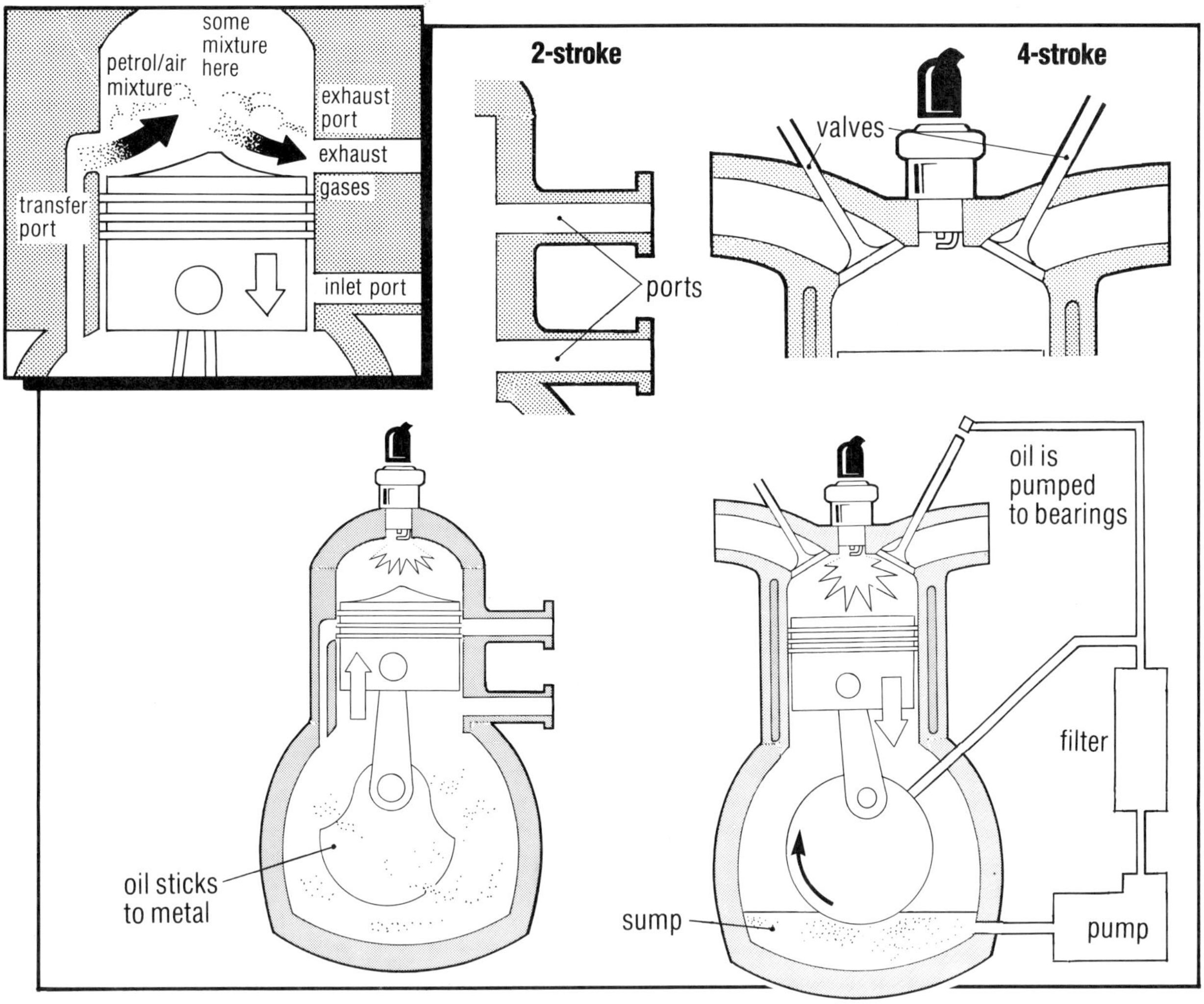

transfer port it forces the exhaust gas out. This means that some of the mixture must mix with some of the exhaust gas and this lowers the efficiency again. The piston crown is sometimes shaped to make the inlet mixture swirl to help push the gases out.

QUESTIONS

1. Make two neat diagrams to show what happens above and below the piston on each of the two strokes of the two-stroke cycle.
2. Explain the advantages and disadvantages of two-stroke engines and four-stroke engines.

6. Diesel and rotary engines

Compression ignition (Diesel) engines

Developed by Rudolph Diesel in 1892; Diesel engines ignite the fuel/air mixture by compression. They work on the principle that when air is compressed into a small space it becomes warm. If diesel fuel is squirted into this hot air it will burst into flames without needing an electric spark.

The four-stroke compression ignition cycle is very similar to the four-stroke spark ignition cycle.

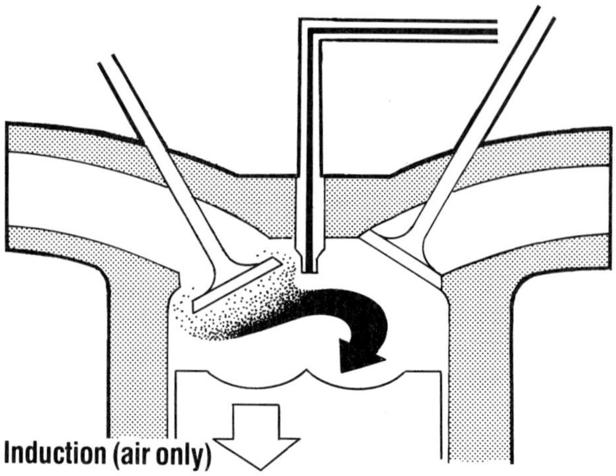

Induction (air only)

INDUCTION
The piston moving down the cylinder draws in air only.

COMPRESSION/INJECTION
The piston moves up the cylinder compressing the air into a very small space. The temperature of the air rises above the 'flash point' of the diesel fuel. Just before the piston reaches 'top dead centre' (t.d.c.) a fine spray of diesel fuel is squirted into the cylinder.

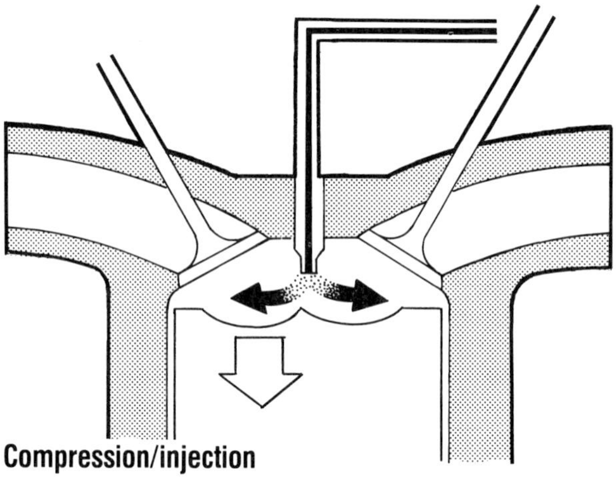

Compression/injection

POWER
The fuel mixes with the hot air and burns very rapidly, expanding and pushing the piston down the cylinder.

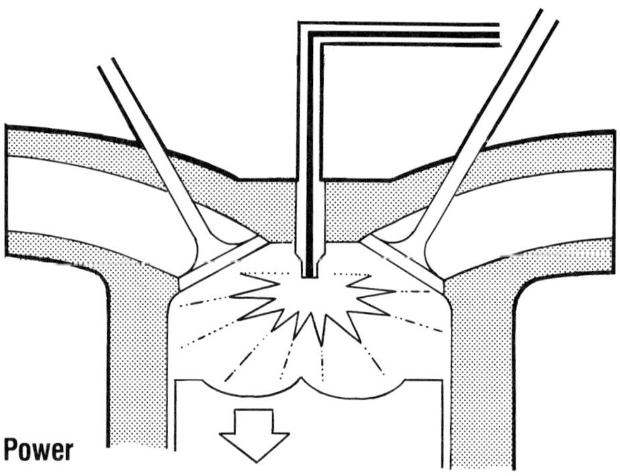

Power

EXHAUST
The piston moves back up the cylinder and forces exhaust gases out.

The fuel is sprayed in through injectors, as a mist of fine droplets which evaporate and burn rapidly as they contact the air at a temperature of around 600 °C. There is a slight delay before burning starts; this is called 'ignition delay' and causes the 'knocking' noise which compression ignition engines make at low speeds.

The compression ratio is approx. 20 : 1 compared with 9 : 1 for spark ignition engines. An accelerator controls the amount of fuel injected and a heater plug (called a *glow plug*) is used to help cold starting.

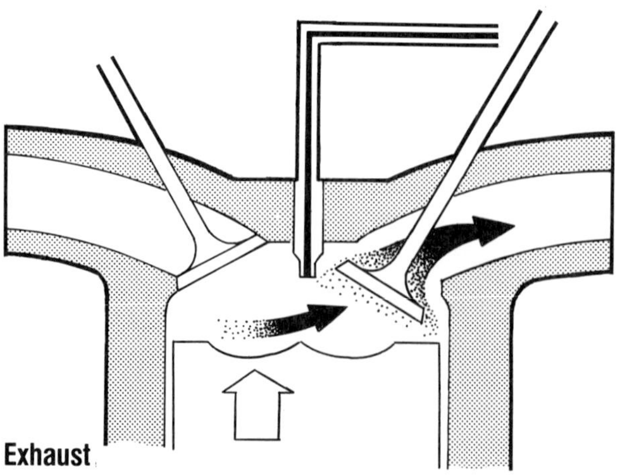

Exhaust

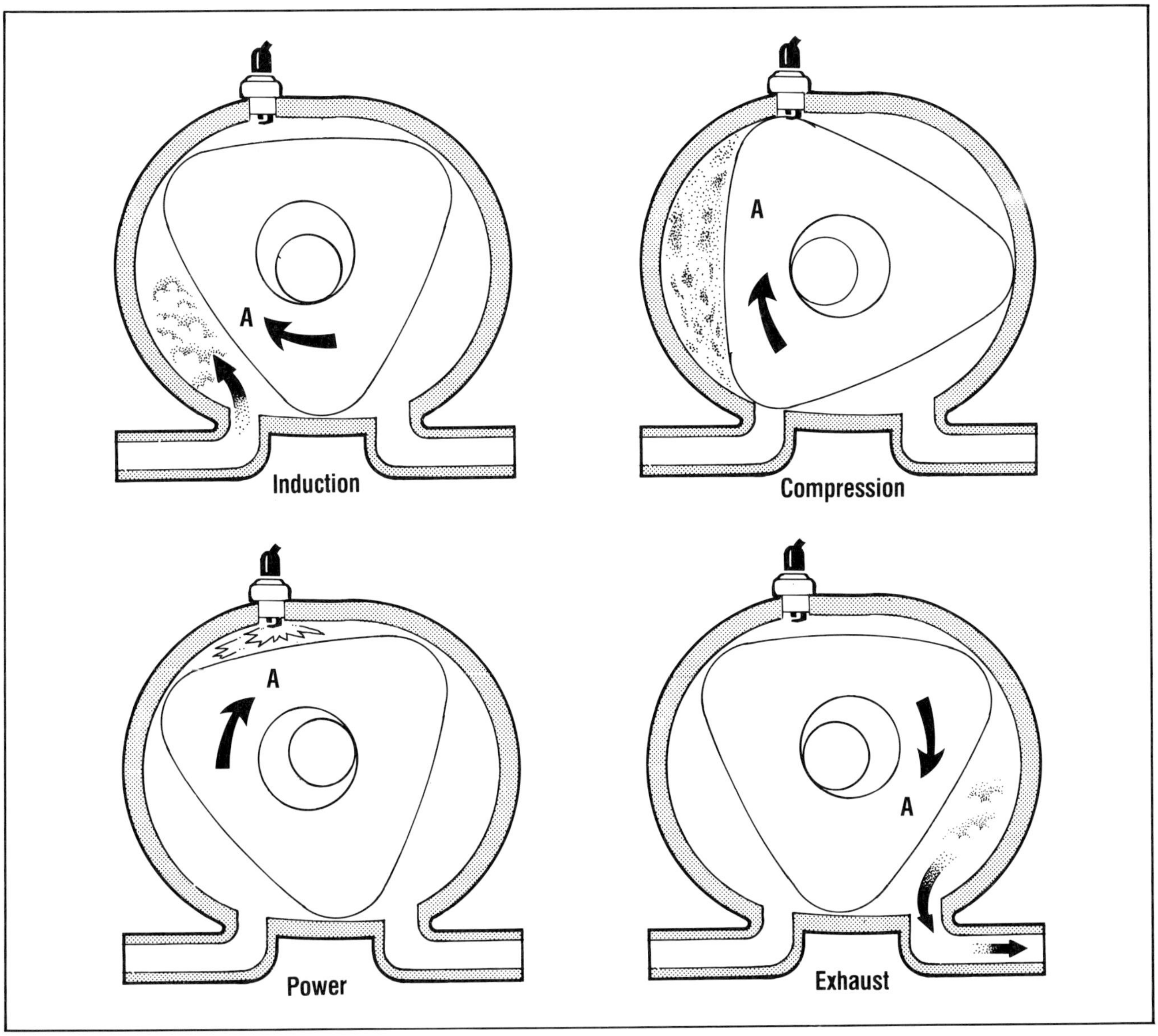

Compression ignition engines are efficient, last longer and are cheaper to maintain than petrol engines but are sometimes expensive, noisier and slower to accelerate. They tend to be used in high mileage vehicles, especially commercial fleets such as taxis, coaches and lorries, where running costs and reliability are important.

Rotary (Wankel) engines

Developed by Felix Wankel, this engine has a triangular rotor instead of a piston. The outer casing is figure-8 shaped. The rotor rotates eccentrically around the output shaft and between the rotor and outer casing are three chambers. Fuel/air mixture is drawn in through an inlet port, compressed between rotor and casing, ignited by a spark plug and as the mixture burns and expands it pushes the rotor around and exhaust gases exit through an exhaust port. This cycle continues on each side of the rotor and produces three power 'strokes' per revolution.

Rotary engines are lighter, smaller and run more smoothly than reciprocating engines but there are problems with fuel consumption and sealing the gap between the rotor and casing.

QUESTIONS

1 Make neat diagrams to explain the four-stroke compression ignition cycle.
2 Write a short essay to explain the advantages and disadvantages of the four-stroke petrol engine, the four-stroke compression ignition engine and the rotary engine.

EXAMINATION QUESTIONS

THE ENGINE

1 The crankshaft is kept rotating smoothly between the power strokes by________ (*NEA*)

2 Pistons have grooves machined around them near the crown. State the purpose of these grooves. (*NEA*)

3 On the induction stroke a mixture of petrol and air enters the cylinder of a 4-stroke engine. What enters the cylinder of a 4-stroke diesel engine on its induction stroke? (*LEA*)

4 An eccentric shape on a shaft for moving another component as the shaft rotates is called ________ (*NEA*)

5 An engine which uses compression of air to produce heat for ignition of injected gas oil is called:
a) wankel;
b) diesel;
c) otto;
d) sterling? (*NEA*)

6 Conventional, modern, four cylinder, in line, four-stroke petrol engines have
a) the firing order 1, 3, 2, 4.
b) coil ignition
c) side valves.
d) four main-bearing crankshaft. (*LEA*)

7 If a twin cylinder four-stroke engine is running at 2000 rev/min how many power strokes will occur each minute? (*LEA*)

8 Draw four simple diagrams to show the four-stroke cycle as used in an OHV petrol engine.
a) Label each sketch with the name of the stroke it is performing.
b) Show clearly on each diagram
(i) the position of the valves, i.e. open or shut.
(ii) the approximate position of the piston.
(iii) direction of travel of piston, i.e. up or down (*LEA*)

9 On the two-stroke engine shown below name the parts numbered 1–6. (*LEA*)

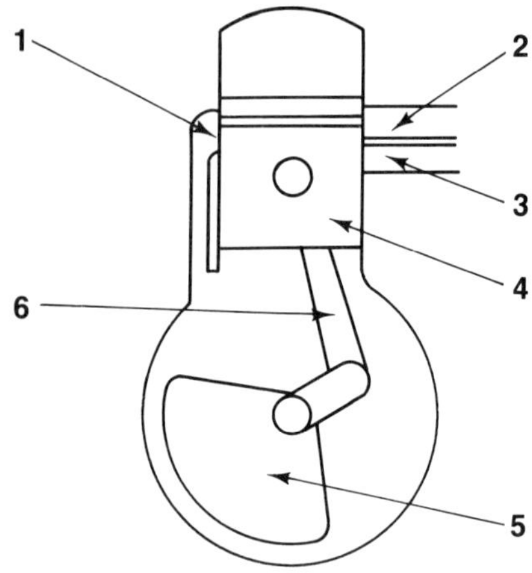

10 a) Name the six lettered components in the diagram below.

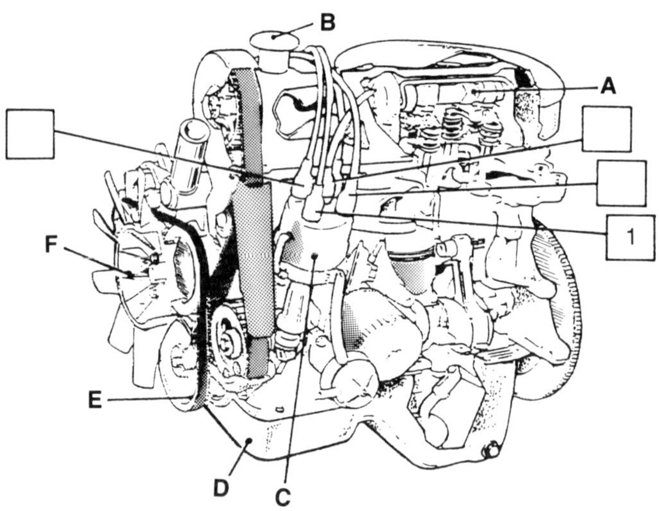

b) The No. 1 H.T. cylinder lead has been indicated on the diagram [1]. Assuming the firing order is 1, 3, 4, 2 and the rotor arm rotates in a clockwise direction, complete the other boxes to indicate the cylinders to which the remaining leads are connected. (*NEA*)

The fuel system

Stores fuel safety before it is pumped to a carburettor where it is mixed with air in the correct proportions for complete combustion in the combustion chamber.

7. Fuel systems

Motor cycles need only a simple gravity flow fuel system as the carburettor is at a lower level than the fuel tank. Cars store petrol below the level of the carburettor and so they use a pump.

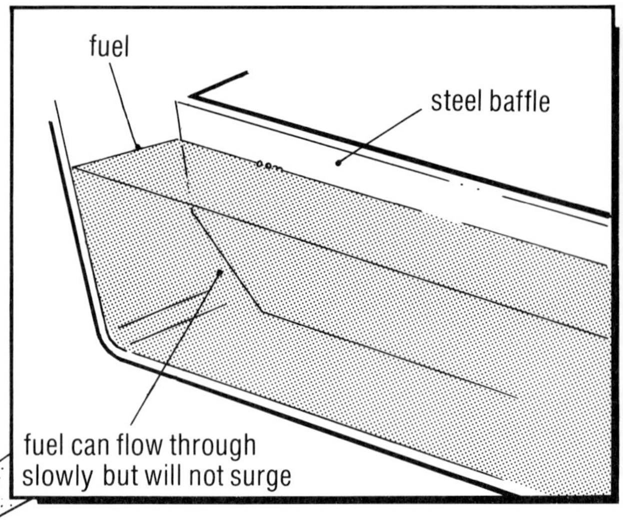

Fuel tank

A pressed steel store for fuel. A filter stops dirt from getting into the carburettor and blocking it. Baffles prevent surge when cornering. A fuel level indicator is also fitted. The fuel filler cap must have a vent hole in it, so that air can enter as fuel leaves the tank.

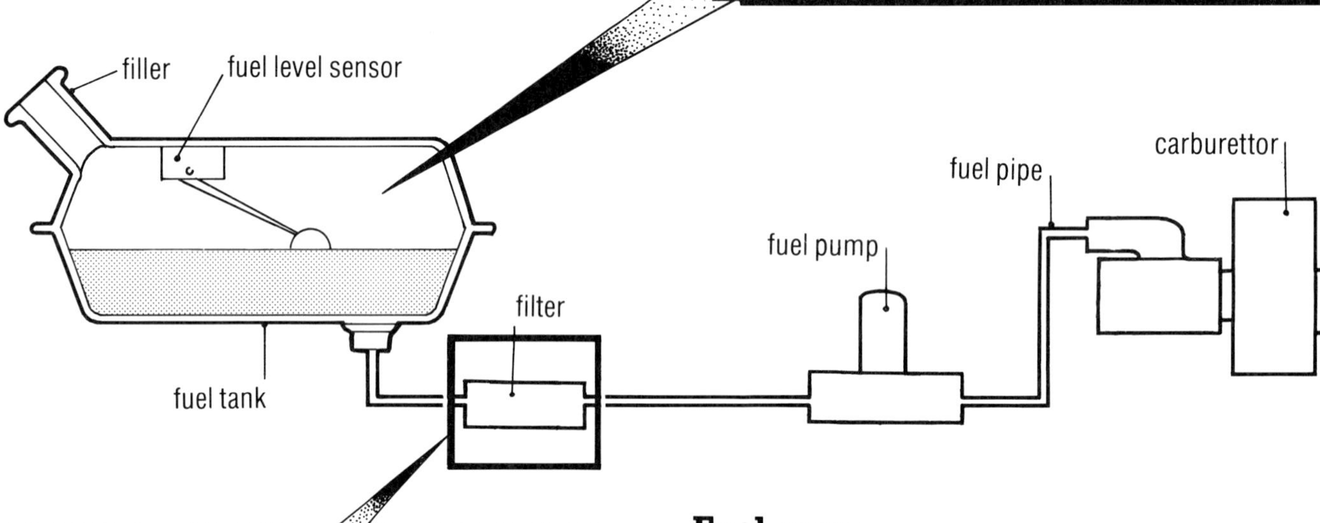

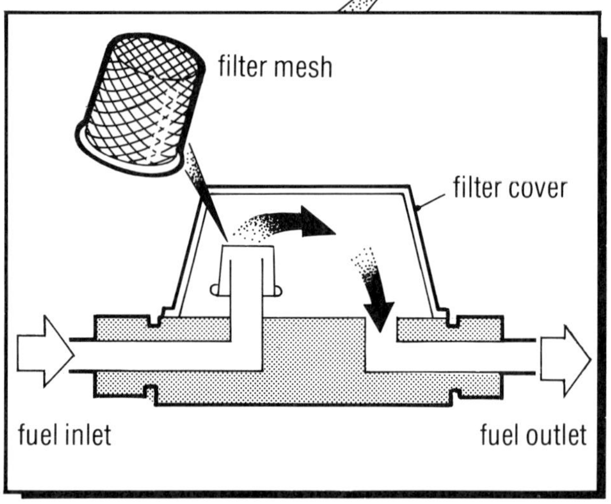

Fuel pipeline

Rigid steel pipes clipped to the bodywork. Flexible plastic pipe for connection to the engine allows for movement.

Fuel pump

This is needed when the fuel tank is lower than the carburettor. Fuel tank is low in cars for safety, stability and space.

There are two types of pump: mechanical and electrical. Mechanical is cheaper and less dangerous in an accident (it stops when the engine stops). Electrical can be positioned anywhere on the car.

MECHANICAL FUEL PUMP

Driven by a cam on the camshaft. The cam pushes a lever which moves a diaphragm. Fuel is drawn in through the inlet valve. When the lever moves back fuel is forced out through the outlet valve.

ELECTRIC PUMP

The electric pump uses a spring-loaded diaphragm. When electricity passes through the solenoid (a coil of wire) a magnetic field is formed. This attracts the arm of the diaphragm. The diaphragm moves down and draws petrol in through the inlet valve. When

Gravity feed fuel system

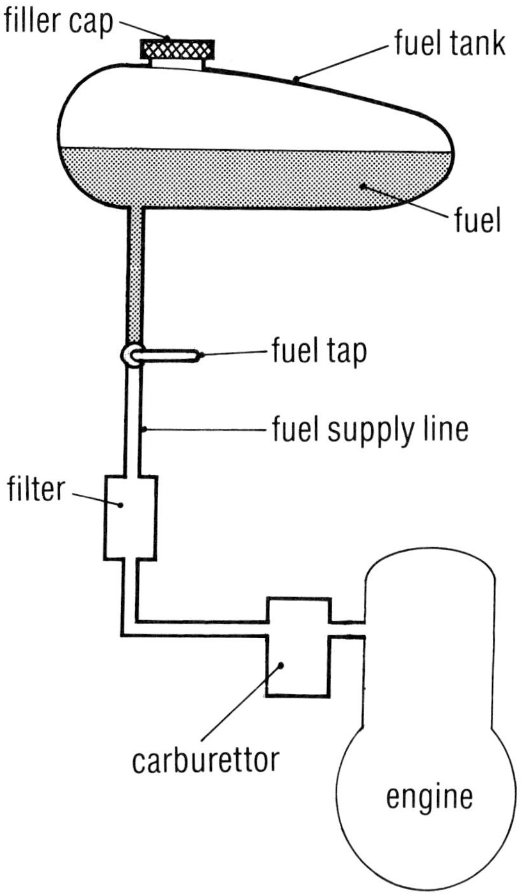

Electric fuel pump

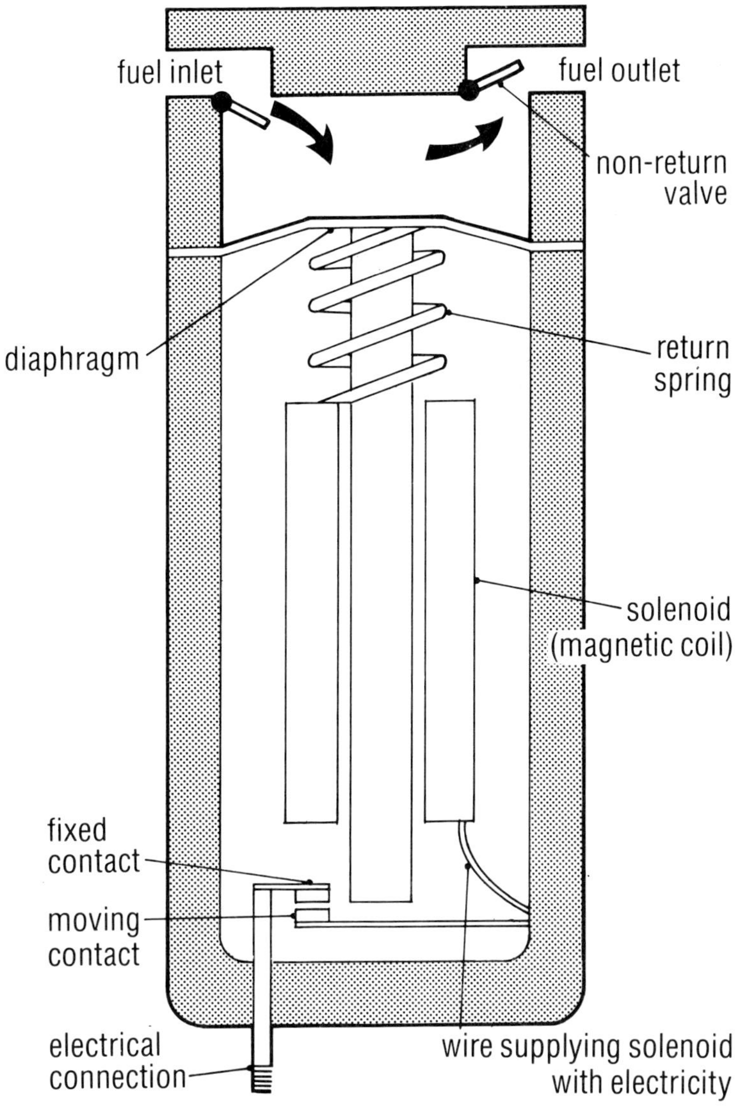

Mechanical fuel pump

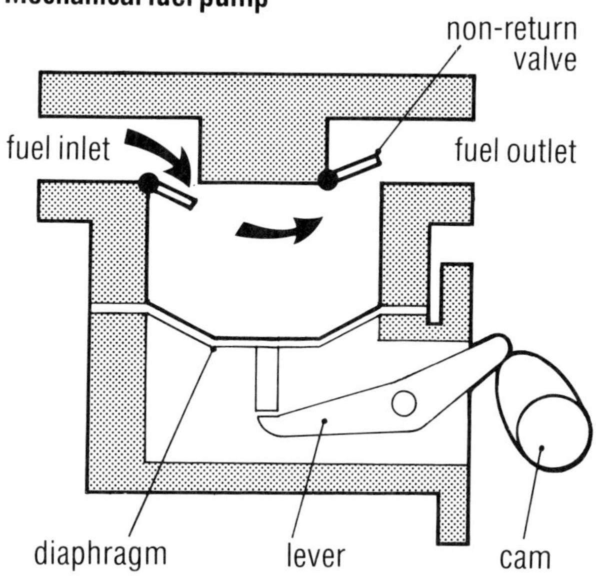

the diaphragm is right down it opens a contact which cuts off the electricity to the solenoid. Because the solenoid has no electricity it loses its magnetism and the spring pushes the diaphagm back up, forcing petrol through the outlet valve. The contact springs close and electricity flows again, attracting the diaphragm, and so on.

QUESTIONS

1 Write two sentences each about each part of the fuel system.
2 Make a neat, labelled diagram of a mechanical fuel pump and write a short explanation of how it works.

Practical: Remove, strip, reassemble and refit a mechanical fuel pump.

8. The carburettor

Petrol alone will not burn, it has to be mixed with air. The carburettor mixes petrol and air into a fine spray which will burn easily and evenly.

PETROL SUPPLY

Petrol is delivered to the carburettor (from the petrol pump) and held in a chamber where its level is controlled by a float – a *float chamber.* As petrol is used the float falls, a needle valve opens and petrol enters. As the petrol level rises the float rises and the needle valve closes, shutting off the supply of petrol. This maintains the level of petrol in the float chamber.

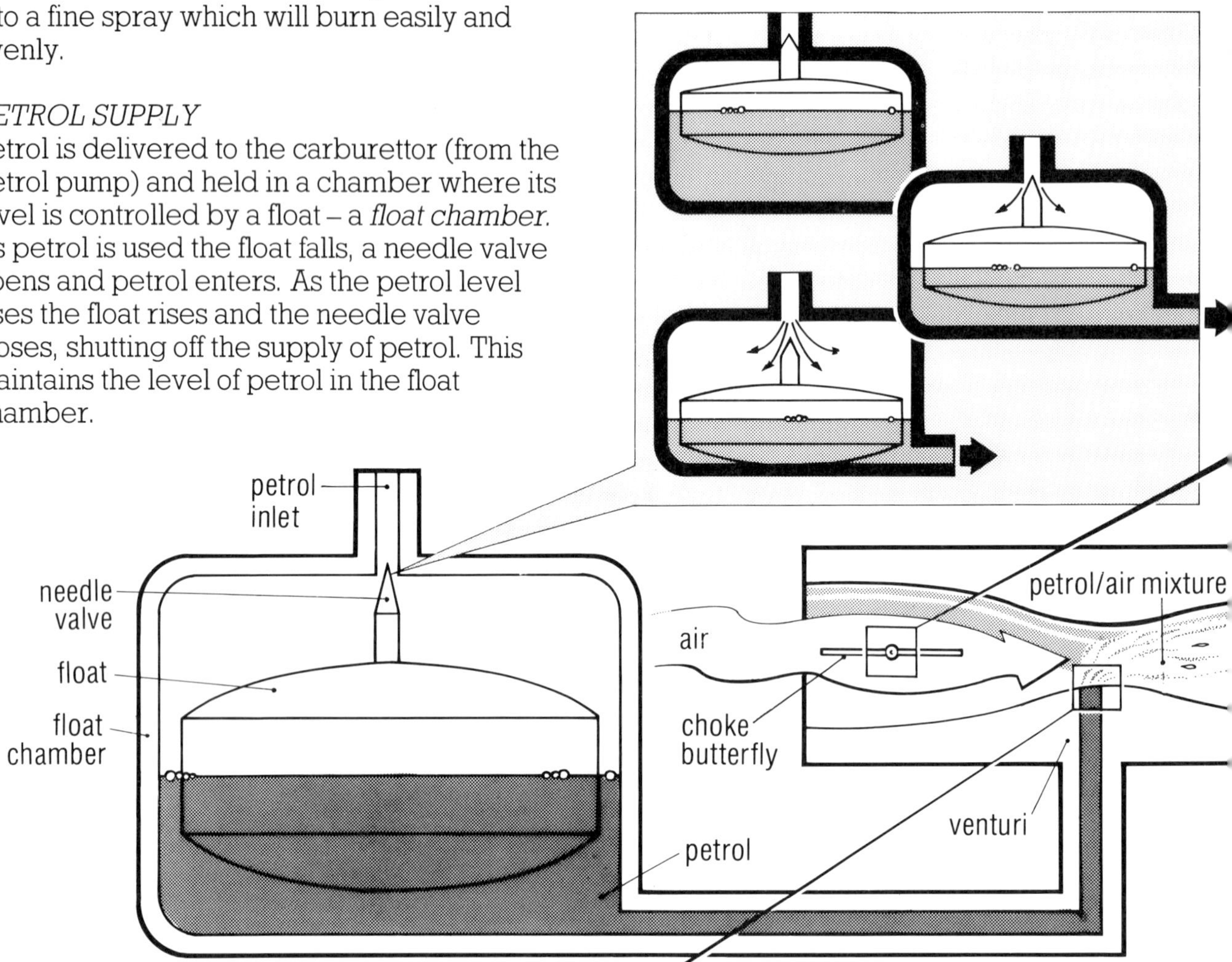

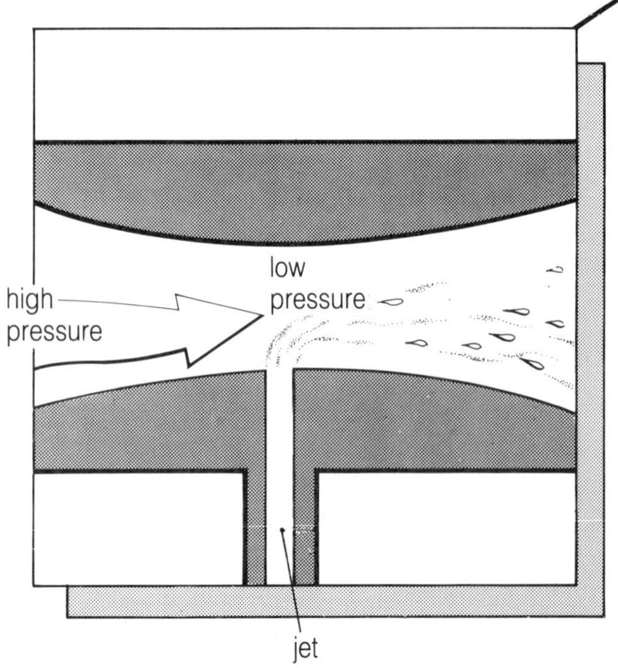

MIXING WITH AIR

If you blow over the end of a tube of liquid you make a spray. This is because the air speeds up and sucks the liquid out of the tube. Speeding up a flow of air reduces its pressure. In the carburettor the inlet is narrowed as it passes the petrol delivery tube to make what is called a *venturi.* As air squeezes through the venturi section it speeds up and petrol flows into the air stream to make a fine spray.

The carburettor mixes 15 parts of air to 1 part of petrol by weight which gives a good mixture for complete burning of the petrol. A throttle is used to control how much of the mixture is passed to the engine – the faster an engine runs the more mixture is needed. The throttle or *accelerator butterfly* is a circular

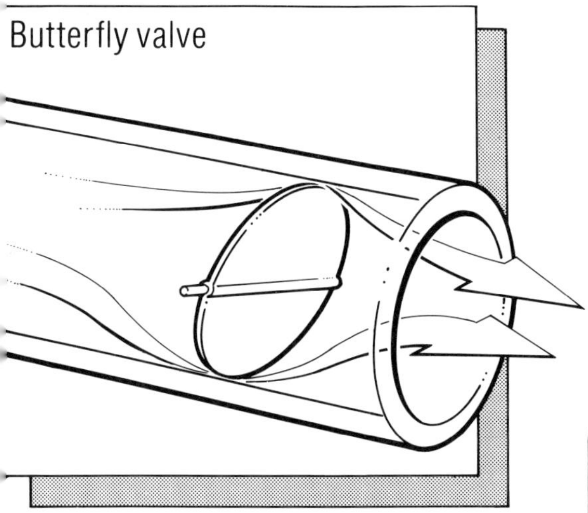

Air filters

Air filters prevent dirt from getting into the carburettor. There are two main types:

1 WIRE MESH
A wire mesh dipped in oil will allow air through but dust will stick to the oil

2 PAPER ELEMENT
Clean and easy to change, the paper catches the dust. A blocked filter reduces airflow and, hence, power. Most filters have a summer and winter setting. In winter air is taken from above the exhaust manifold which helps cold running but gives slightly less power.

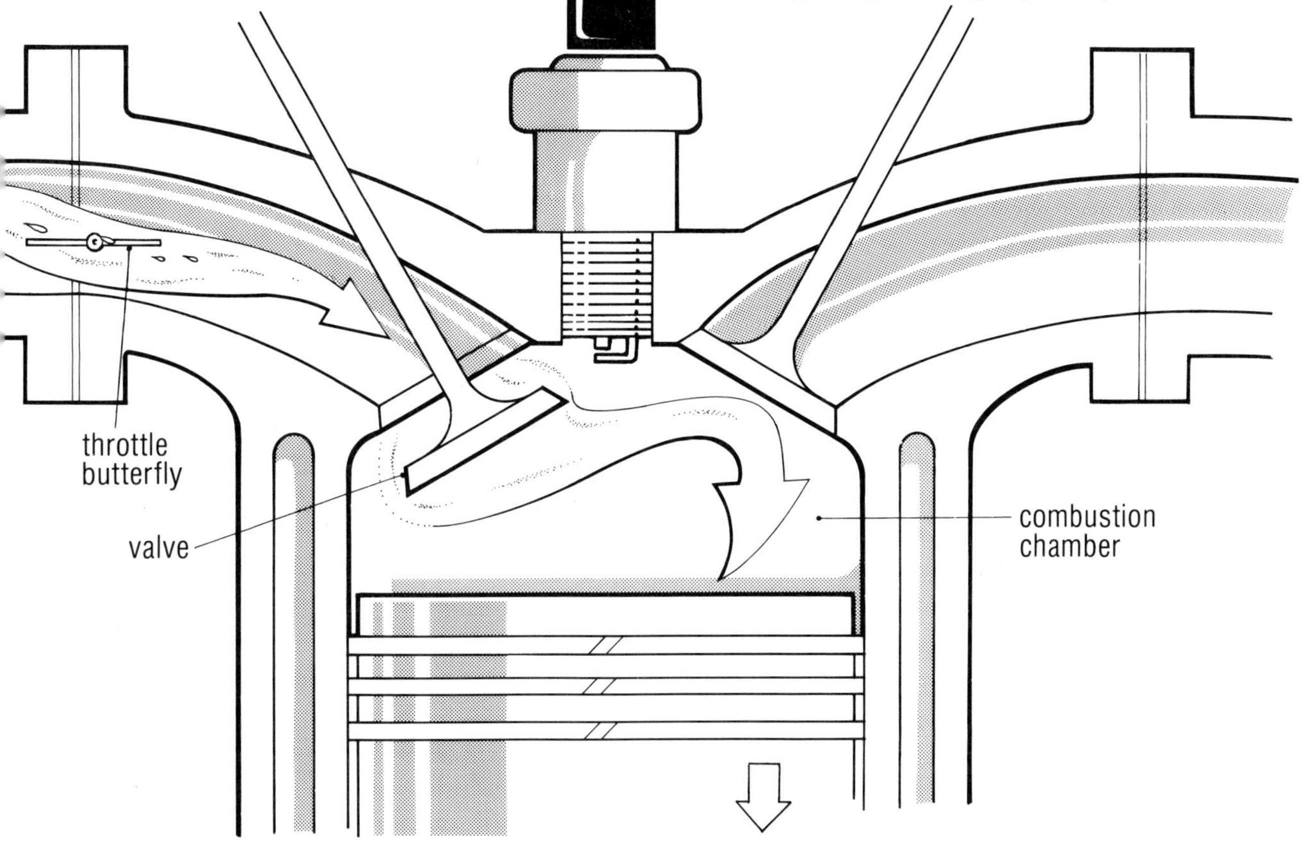

steel disc, pivoted in the centre. Another butterfly valve is the *choke butterfly* and this reduces the amount of air that can pass so that the petrol/air mixture becomes 'richer' (more petrol to each part of air than normal). The choke is used to start the engine from cold when petrol does not easily turn into a vapour but tends to condense and become liquid again on the cold metal surfaces.

The throttle butterfly is positioned between venturi and cylinder, the choke butterfly is before the venturi.

QUESTIONS

1 Make a neat, labelled diagram of a section through a carburettor.
2 Write an explanation of how a simple carburettor works; mention how the float chamber maintains a constant fuel level and what happens in the venturi.

Practical: Strip, examine, clean and reassemble a fixed jet carburettor.

9. Fixed and variable jet carburettors

The normal air/fuel ratio is 15 : 1 by weight. If this were the only ratio needed then the carburettor could be very simple. However, the air/fuel ratio changes. Acceleration uses more fuel and the ratio may be 12 : 1; starting from cold uses a lot more fuel and the carburettor has to supply a 2 : 1 mixture which may drop to 16 : 1 for cruising speeds with a warm engine. To deliver the correct mixture strength for all of these running conditions the carburettor needs either a variable jet or several different fixed jets.

Fixed jet carburettor

Sometimes called a variable depression carburettor, this type uses different jets for idle, normal running and acceleration.

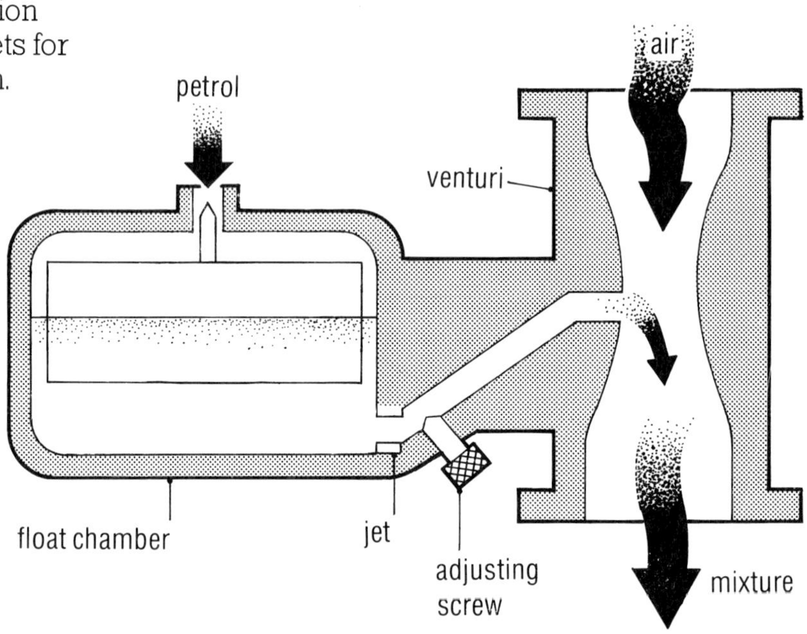

1 SLOW RUNNING JET
Supplies a small amount of fuel for idling speed (which may be adjusted using an adjusting screw).

2 MAIN JET
Supplies the mixture for normal running. As the throttle is opened more air flows. Some air flows through mixing tubes and bubbles out of small holes into the petrol making an 'emulsion' which goes to the venturi where it is mixed with the main air flow.

3 ACCELERATOR PUMP
Gives a short 'squirt' of petrol when the throttle is pressed down. Uses more fuel but gives rapid acceleration.

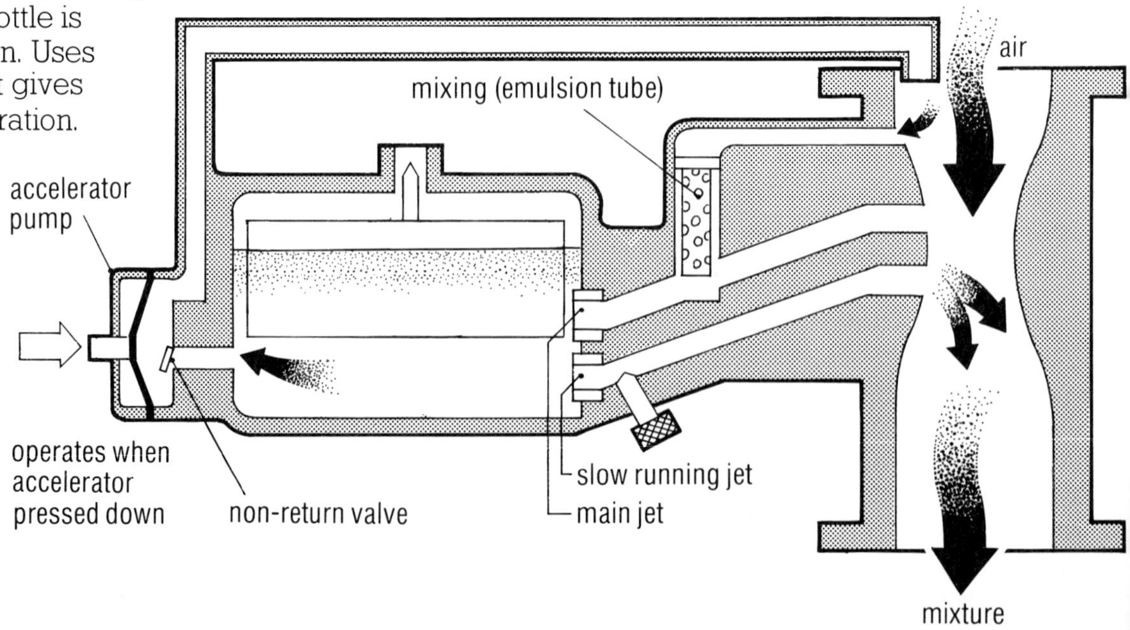

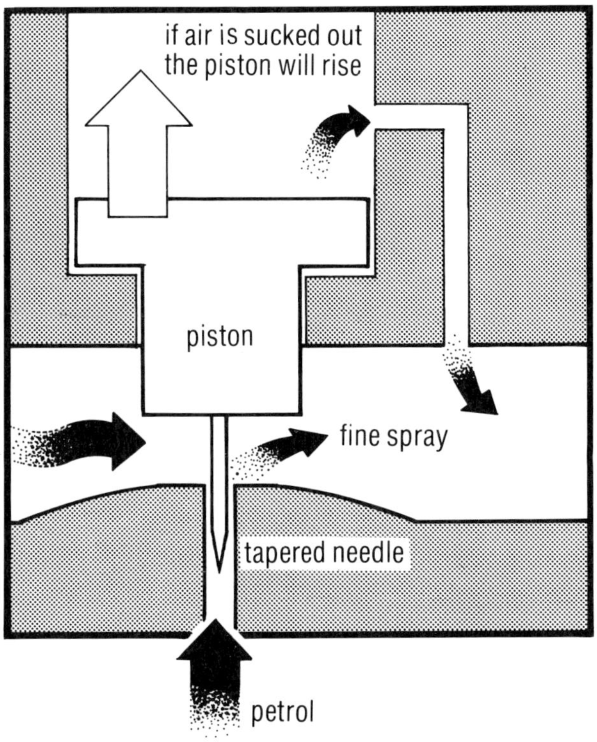

Variable jet carburettor

Sometimes called a constant vacuum carburettor, this type uses changes in airflow to move a piston and tapered needle in a jet. This alters the effective size of the jet.

If air is sucked out of the space above the piston it will rise. As it rises more petrol is drawn into the airstream.

For cold starting the jet can be lowered to give an extra rich mixture.

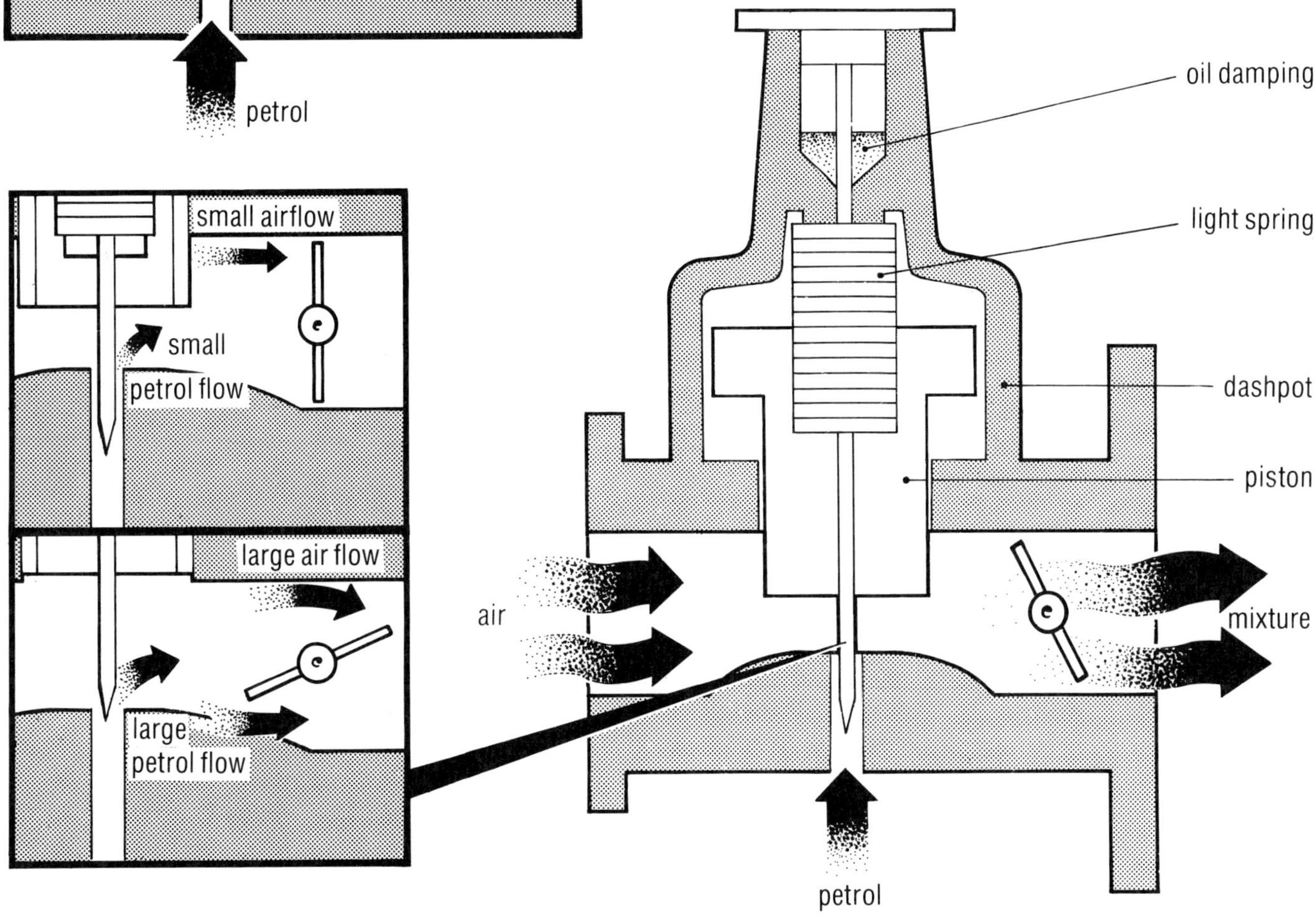

QUESTIONS

1 Why is a simple carburettor not suitable for use in a car?

2 Make neat, labelled diagrams to explain the differences between fixed jet and variable jet carburettors.

Practical: Remove a variable jet carburettor from an engine. Strip, clean, reassemble and refit. Adjust slow running.

10. Petrol

Petrol engines are sometimes called 'heat engines'. They work by changing the *chemical* energy which is stored in petrol into *heat* energy and then changing this heat energy into *mechanical* (moving) energy.

As the fuel/air mixture burns heat energy is produced. This makes the combustion gases expand and push the piston down. Heat energy is changed to mechanical energy. The piston moves up and down in a 'reciprocating' movement. The connecting rod and crankshaft change reciprocating movement into rotating movement.

Properties of petrol

INDUCTION
Petrol has to be *volatile* (i.e. turn into a vapou easily).

COMPRESSION
Petrol must not detonate on compression, it must have a high *octane rating.*

POWER
Petrol must burn well.

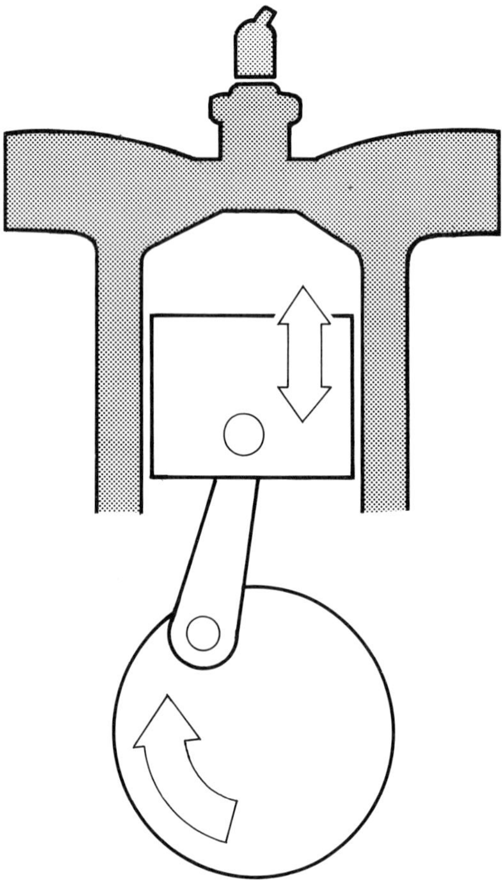

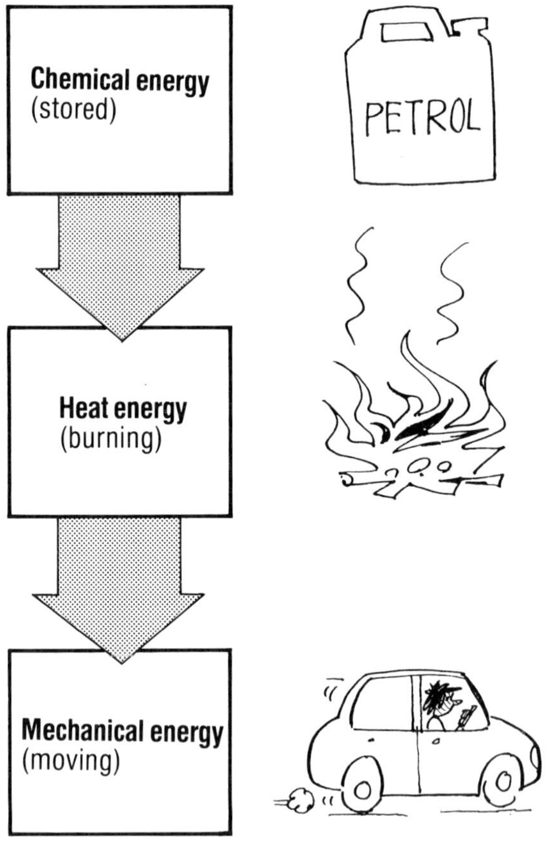

EXHAUST
Petrol is a hydrocarbon fuel – it contains hydrogen and carbon. When it burns the hydrogen combines with oxygen to form water vapour and the carbon combines with oxygen to form carbon dioxide and carbon monoxide. Additives contain lead which combines with oxygen to form lead oxides which form the third constituent of exhaust gases.

Petrol + Air ⟶ **Exhaust**

Hydrogen + Oxygen ⟶ **H_2O (water)**

Carbon + Oxygen ⟶ **CO_2 (carbon dioxide)**
⟶ **CO (carbon monoxide)**

Lead + Oxygen ⟶ **PbO (lead oxide)**

Properties of Petrol

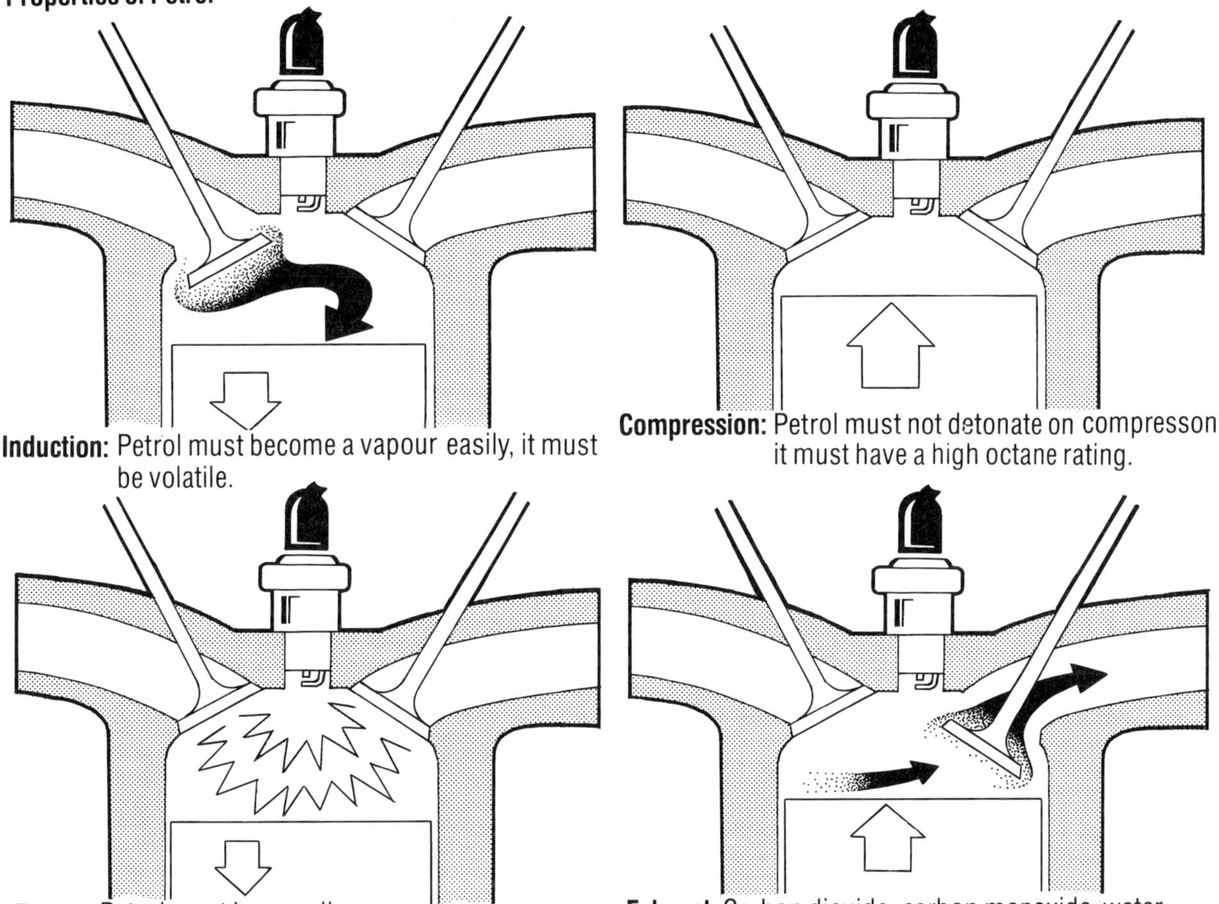

Induction: Petrol must become a vapour easily, it must be volatile.

Compression: Petrol must not detonate on compresson, it must have a high octane rating.

Power: Petrol must burn well.

Exhaust: Carbon dioxide, carbon monoxide, water vapour and lead oxide are left when petrol burns.

Octane number

★★ 2-star is 90 octane
★★★ 3-star is 94 octane
★★★★ 4-star is 97 octane

The higher the octane number the better the fuel. High compression ratio engines need high-octane fuels. A low-octane fuel in a high compression engine will not burn smoothly but will detonate causing 'knocking'. (A high-octane fuel in a low compression engine will not improve performance.)

Additives

Some extras are added to petrol to improve it, make it burn better and leave fewer deposits in the cylinder:

1 Dyes to identify different grades.
2 Ignition control additives to improve the way it burns.
3 Anti-knock additives to increase its octane rating.

QUESTIONS

1 Make a neat, labelled diagram of a section through a cylinder and write a note about the energy changes that take place inside the engine.

2 Write two or three sentences about each of the following: octane number, additives, hydrocarbon fuel, properties of petrol.

11. Combustion

The space between the piston and cylinder head where the mixture burns is called the *combustion chamber*. This space must be small so that it stays hot and smooth to enable the flame to spread easily and quickly through the mixture. There are several types of combustion chambers as shown below.

During combustion hydrogen atoms combine with oxygen atoms to form water vapour and carbon atoms with oxygen atoms to form carbon dioxide (or carbon monoxide if combustion is not complete). The ideal mixture for combustion is 1 part fuel to 15 parts air by weight. If the fuel and air is not thoroughly mixed burning will not be even and power will be lost. Too much petrol will give a rich mixture which gives more power but wastes petrol. Too little petrol gives a weak mixture and less power but more efficient combustion and lower fuel consumption. (If the mixture is too weak the engine will overheat.)

About 25°–30° before t.d.c. a spark occurs. There is a very slight 'ignition lag' (about $\frac{1}{600}$th of a second) before the mixture starts to burn. Then the flame front moves very quickly (20 m/s) through the charge; it moves quickly but smoothly. If the charge burns instantly throughout, it produces a shock wave. This is called 'detonation' and makes a 'knocking' or 'pinking' sound as the shock wave hits the cylinder wall or piston crown. Detonation causes overheating, damage to pistons or engine failure. The rapid increase and decrease in pressure causes power loss.

CAUSES OF DETONATION

- Poor fuel, octane number too low for the compression ratio of the engine.
- Advanced ignition or incorrect plug.
- Weak mixture causes hot engine running.
- High speeds, heavy loads, carbon, overheating.

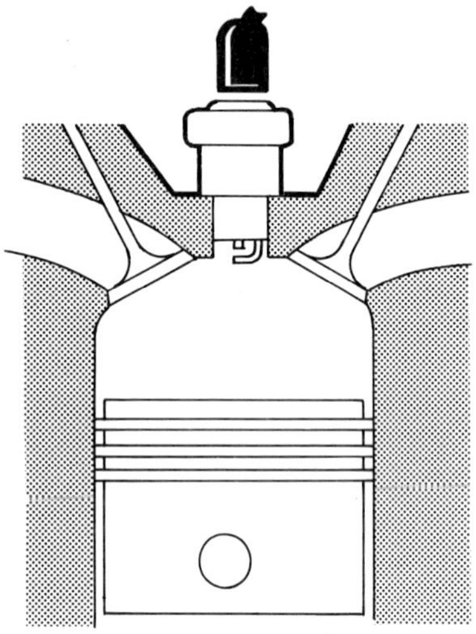

Hemispherical

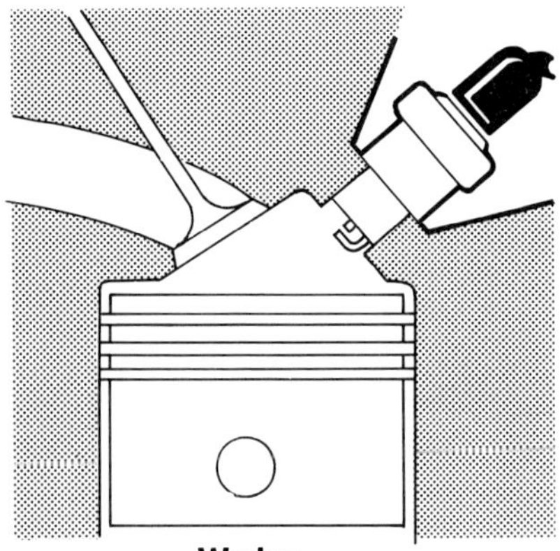

Wedge

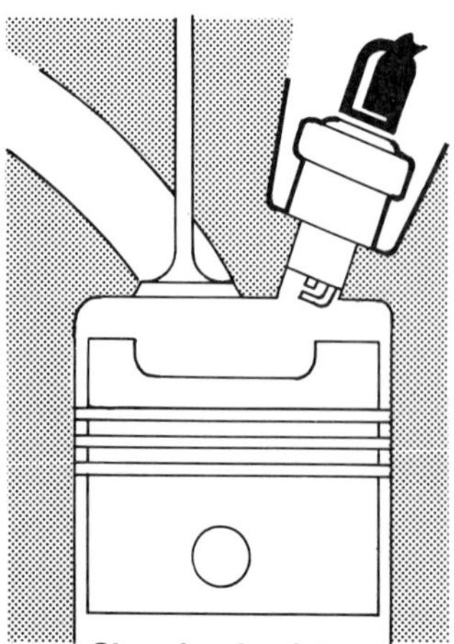

Chamber in piston

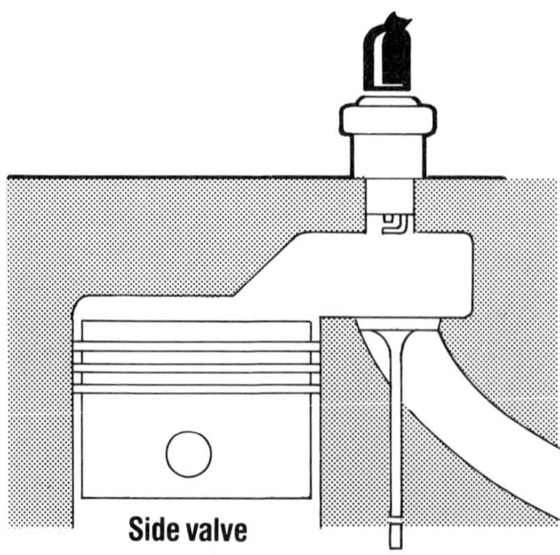

Side valve

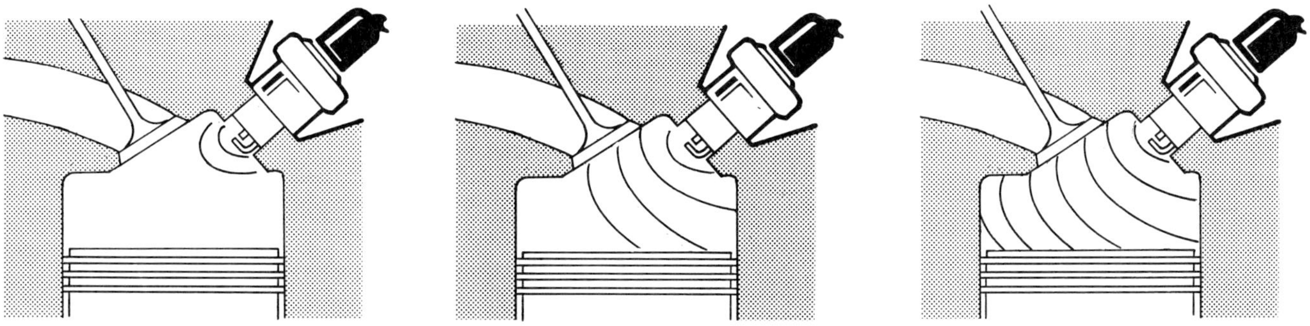

Correct Burning

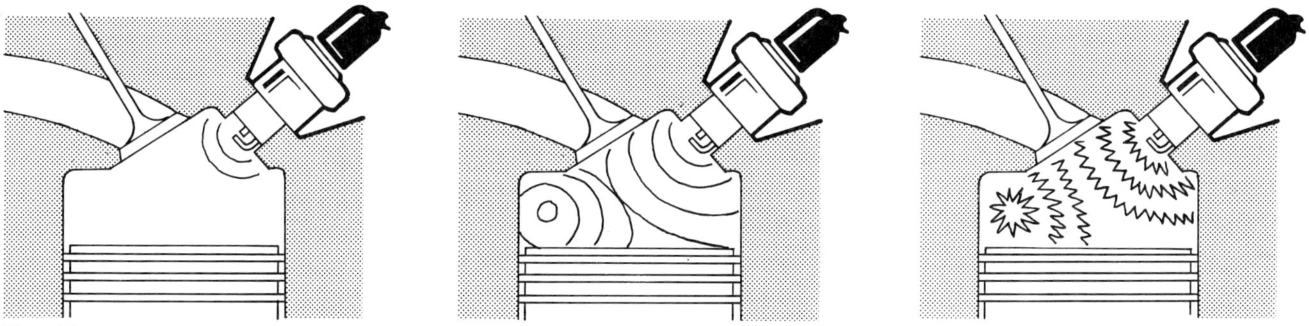

Knocking

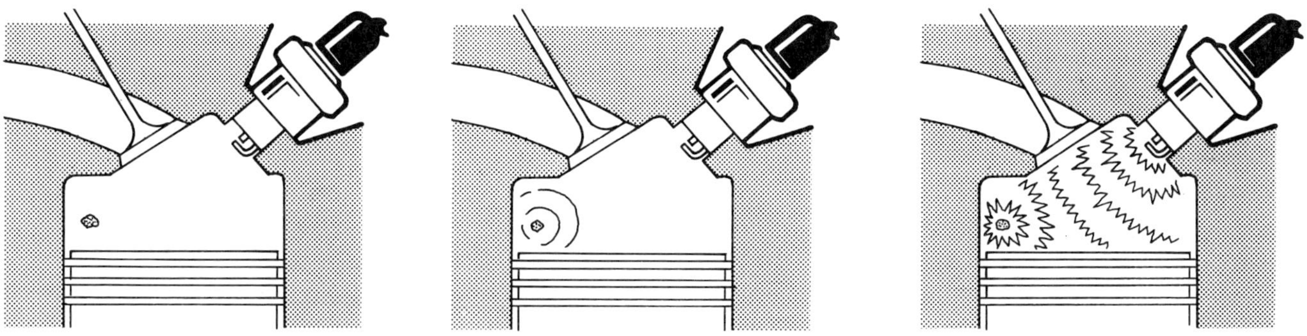

Pinking

Correct burning

The spark plug ignites the mixture and the flame front moves very quickly and smoothly across the piston.

Knocking

Low-octane petrol may cause a high temperature which ignites part of the mixture before the flame front reaches it. When they meet there is a violent 'knock'.

Pinking

Carbon in the combustion chamber may become heated and ignite part of the mixture before the plug sparks. A high pitched 'pinking' noise is caused when the two flame fronts meet.

Pre-ignition

Hot spots in the engine ignite the mixture before the spark. The mixture burns earlier in compression stroke, high pressure opposes piston and power is lost. The engine also tends to 'run on' when the ignition is switched off.

QUESTIONS

1. Make diagrams and short notes about the different types of combustion chamber.
2. Describe what is meant by correct burning, knocking and pinking.

Practical: Remove cylinder head, 'decoke' combustion chamber, fit new gasket and replace cylinder head.

EXAMINATION QUESTIONS

THE FUEL SYSTEM

1 Why is petrol mixed with air before combustion?

2 Name two basic types of carburettor. (*LEA*)

3 Petrol is a hydrocarbon fuel. What is a hydrocarbon?

4 What is a typical fuel air ratio for:
a) cold starting
b) normal running
c) acceleration?

5 Which one of the following components could form part of a mechanically operated petrol pump?
a) float,
b) rotor,
c) jet,
d) diaphragm. (*LEA*)

6 Make a neat, labelled sketch of the fuel system of a car.

7 The throttle valve, which is a metal disc on a spindle in the carburettor, is called ________. (*NEA*)

8 The drawing below shows a carburettor in section. Name the parts numbered 1–6. (*LEA*)

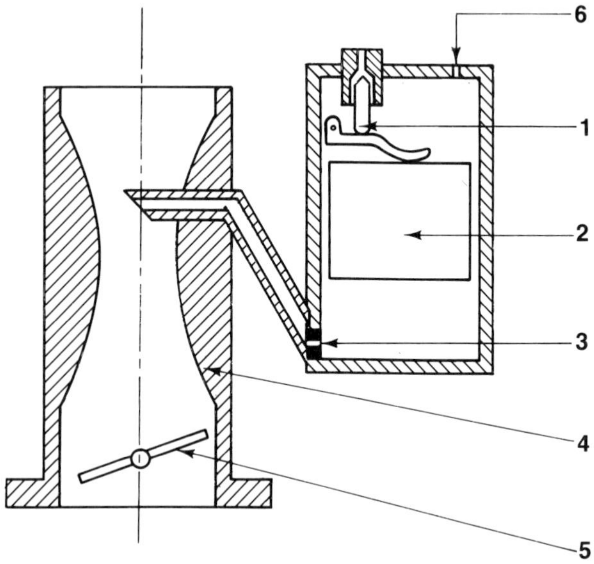

9 What could happen if the vent on a petrol tank became blocked on a long journey? (*NEA*)

10 Make a neat, labelled diagram of a simple carburettor and explain how it works.

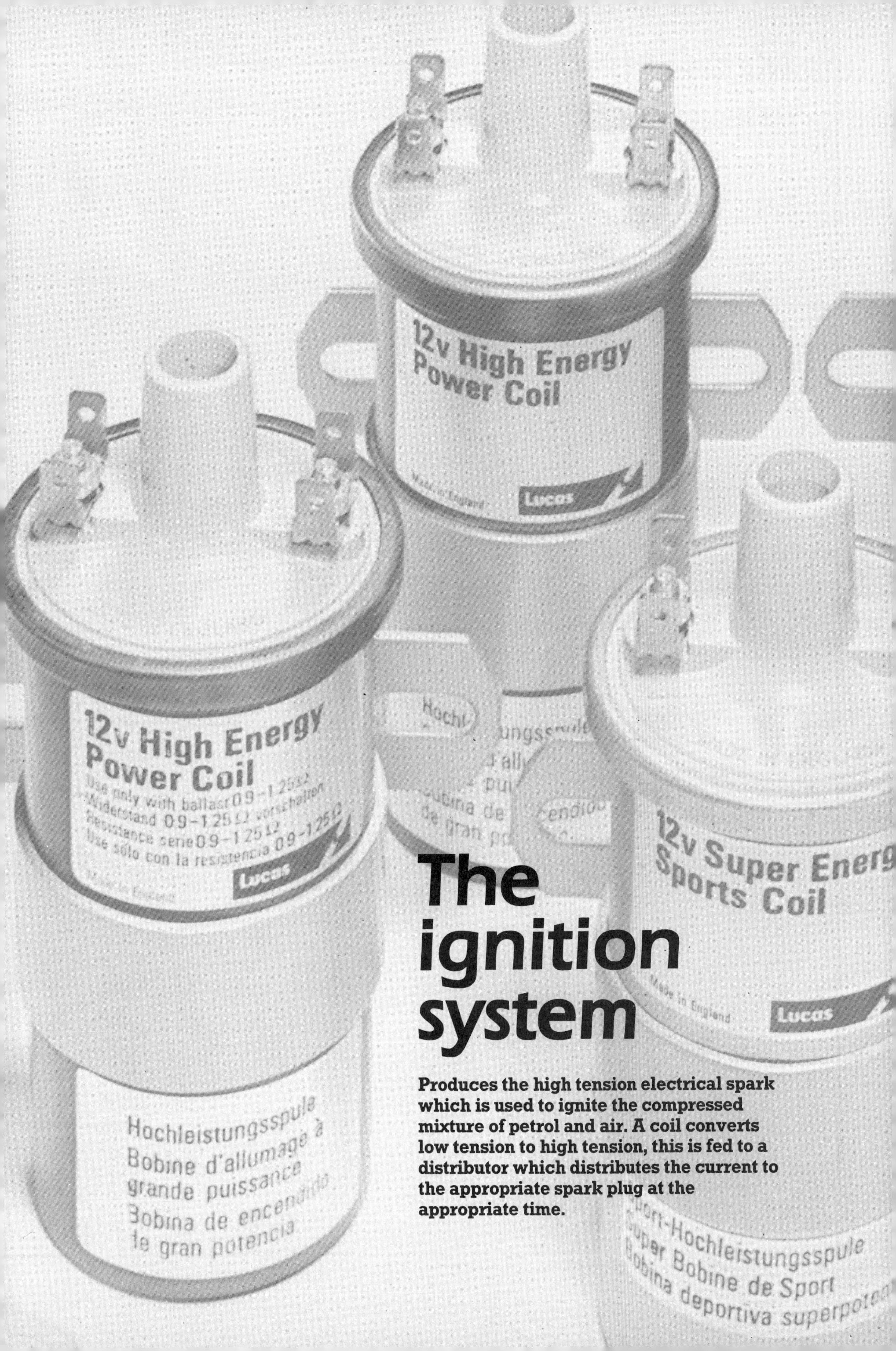

The ignition system

Produces the high tension electrical spark which is used to ignite the compressed mixture of petrol and air. A coil converts low tension to high tension, this is fed to a distributor which distributes the current to the appropriate spark plug at the appropriate time.

12. Spark plugs

The diagram (right) shows what happens inside a spark plug. A wire (called an *electrode*) passes through the centre of the plug. There is a small gap between this and another electrode. When both electrodes are connected to the battery there will be a spark across the spark plug gap.

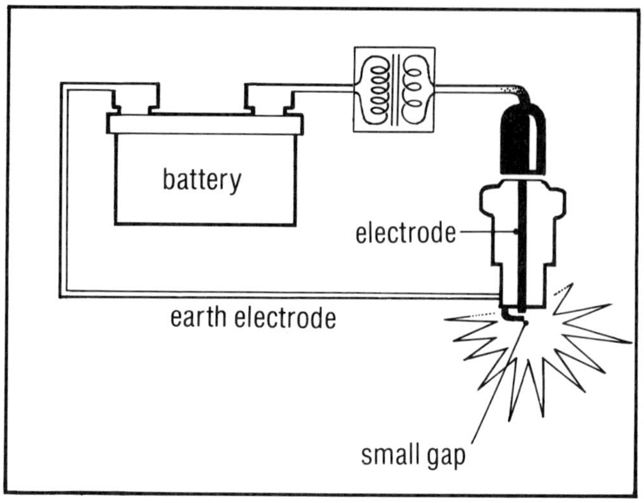

The main parts

The metal body of the plug (the *shell*) holds a *ceramic insulator* which makes sure that the electricity does not leak but flows along the *central electrode*. High tension (HT) current flows down the central electrode and jumps the *spark plug gap* to the *earth electrode*. The spark ignites the petrol/air mixture.

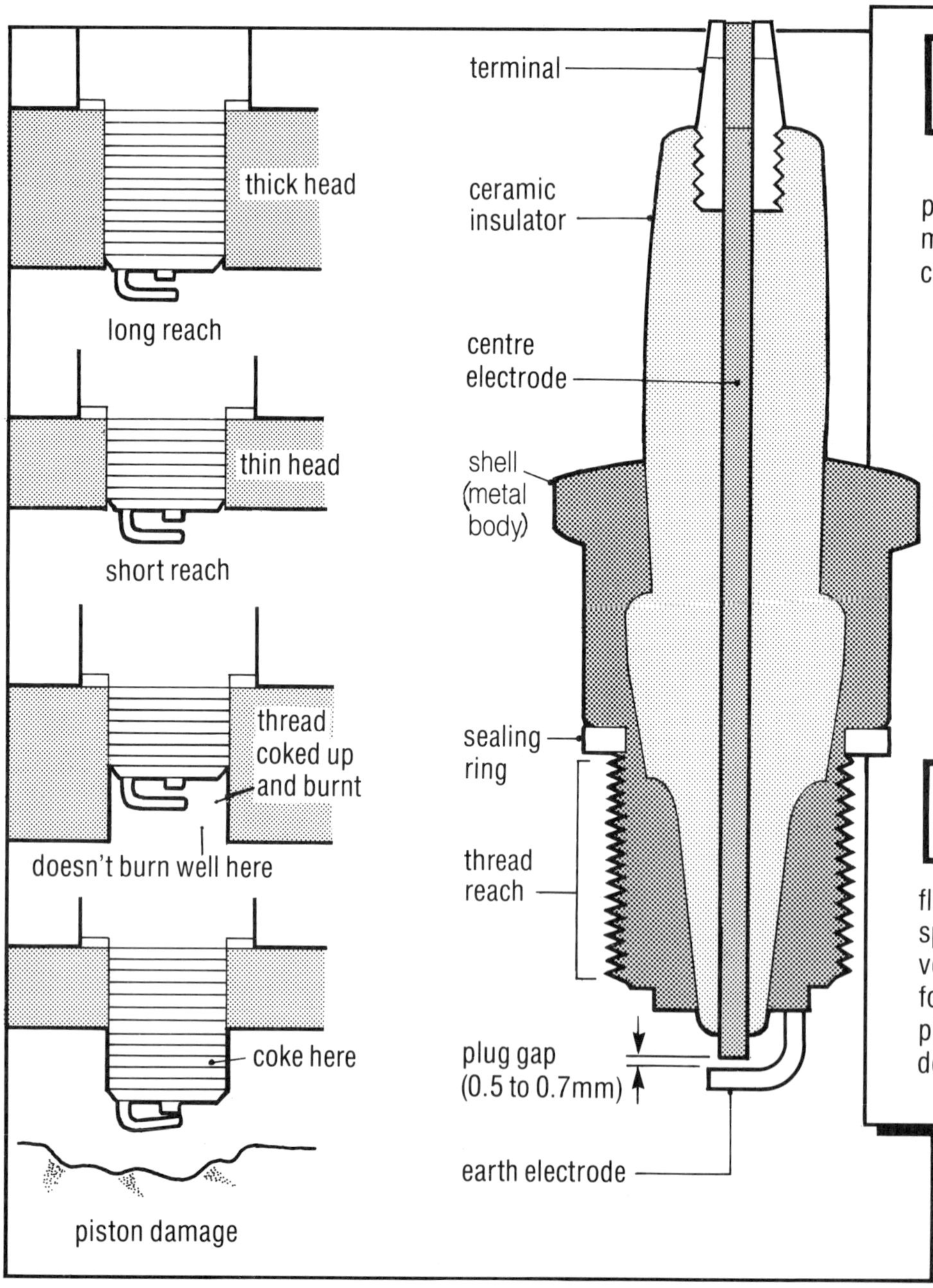

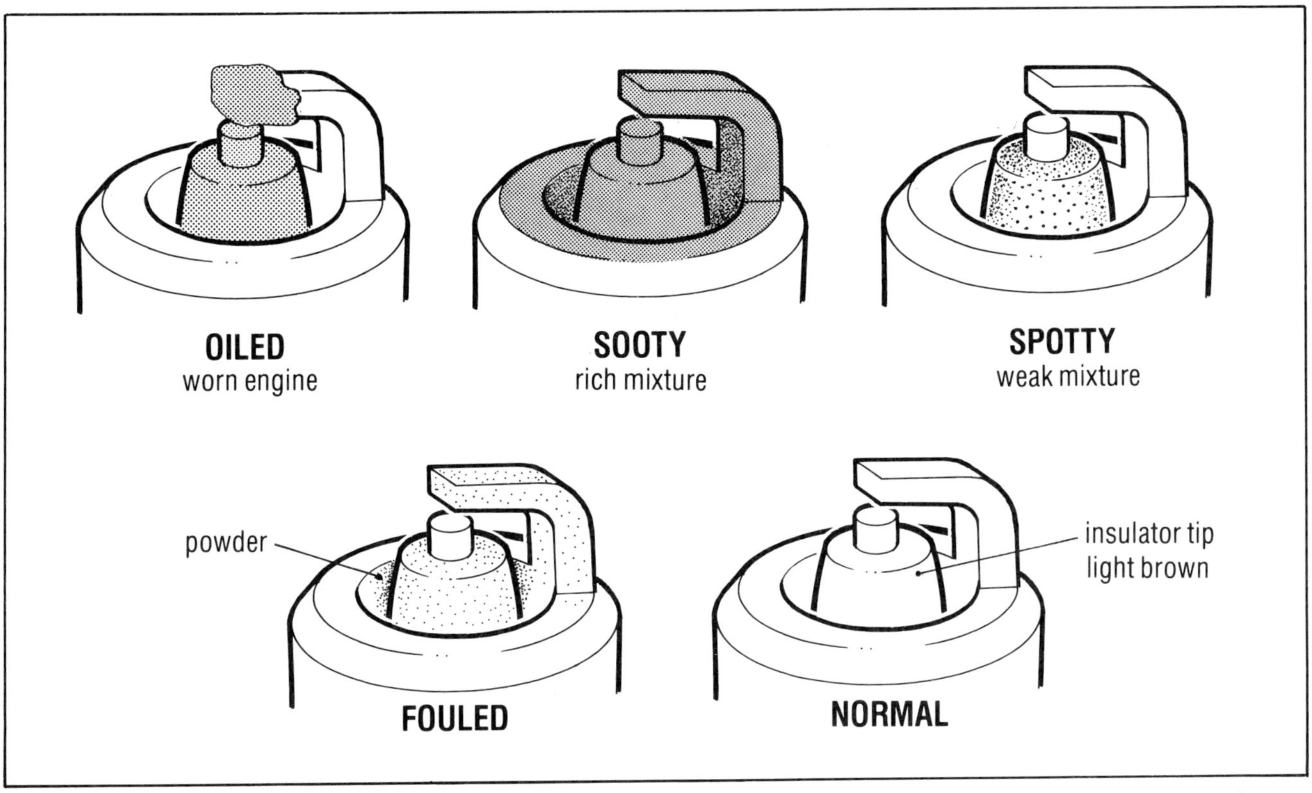

Short reach and long reach

The threaded part of the plug is called the 'reach'. A long reach plug has a long thread, a short reach plug a short thread! The reach does not affect the spark. But what if you use a short reach plug in a long reach head or vice versa?

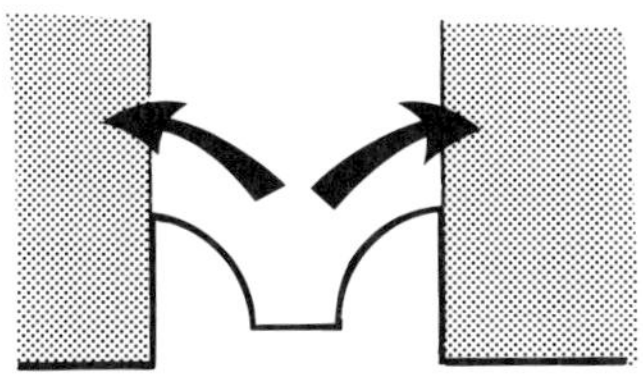

COLD
heat flows
out quickly

Spark plug condition

Because the mixture burns around the spark plug you can find out about the engine running conditions by looking at the plugs.

HOT
heat travels
further plug
stays hot

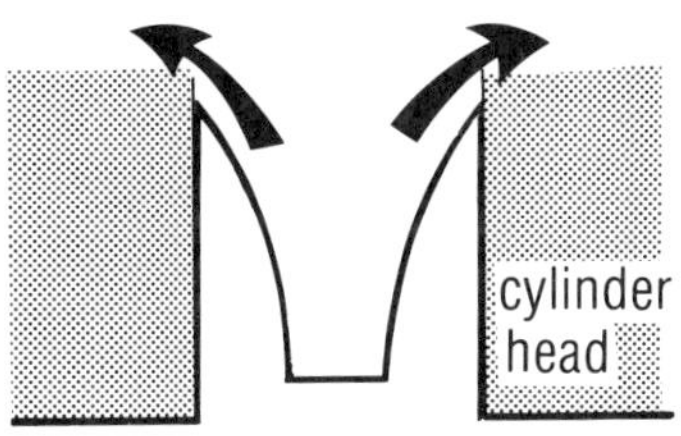

Hot and cold

Producing over 100 sparks per second makes the spark plug very hot. Heat travels from the electrode, through the insulator, to the shell and into the cylinder head. The diagram on the right shows the shape of the insulator and the path of the heat. A high performance engine needs to remove heat very quickly or it will overheat – it needs a *cold plug*. A slow running engine will run cool and therefore needs a *hot plug*. It is important to use a plug with the correct heat range for the engine. Using a hot plug in a high performance engine will cause overheating, a cold plug in a cold engine will cause oiling or fouling of the spark plug.

QUESTIONS

1 Write a short note about plug condition.
2 Make a neat, labelled sketch of a spark plug and write a short explanation of the purpose of each part.

Practical: Remove, examine, clean, reset and replace a spark plug.

13. The ignition coil

The ignition coil is actually *two* coils of wire. The first coil has a few hundred turns of thick wire, the second has thousands of turns of thin wire. A 12-volt current is passed through the first (primary) coil. When this is switched off for a split second a current of 12 000 volts or more is induced in the second (secondary) coil. This 'high-tension' current passes along the HT lead to the spark plug.

How does a coil work?

When electricity passes through a wire, the wire becomes magnetic. This can be shown by putting a wire through a piece of card and sprinkling iron filings on the surface.
If the wire is connected to a battery and the card is tapped the filings will move and show the shape of the magnetic field.

If the wire is coiled the magnetism is more concentrated and powerful. To strengthen the magnetism even more the wire can be wrapped around a piece of soft iron. The soft iron core concentrates the magnetic field.
When the electricity is switched off the magnetic field collapses.
When a magnet is moved past a wire a short burst of electricity is formed. This is called 'electro-magnetic induction'.
A magnet moving past a coil of wire will 'induce' a voltage.
In the ignition coil, as a magnetic field collapses the lines of force pass through the secondary coil and this creates a high voltage.

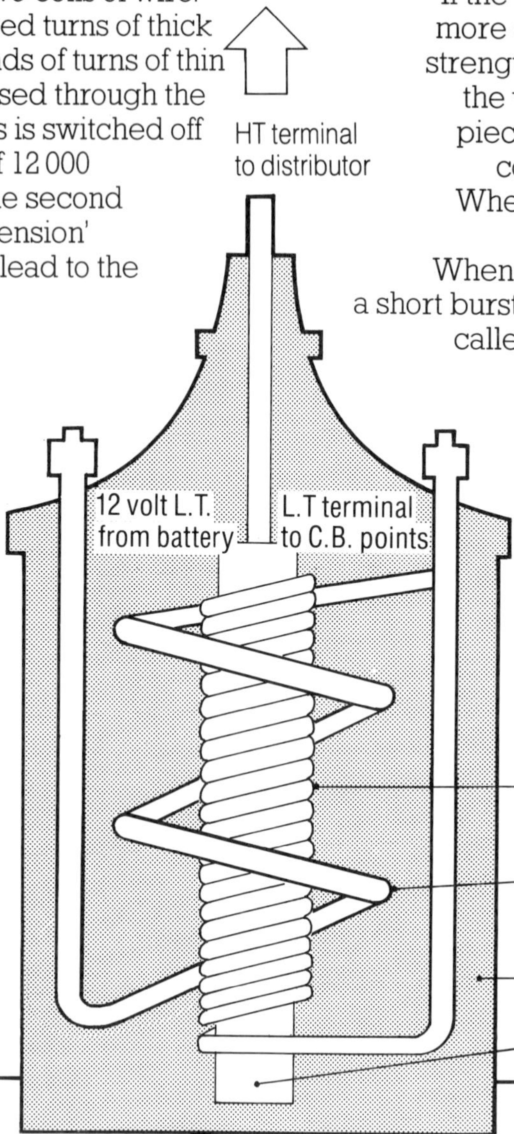

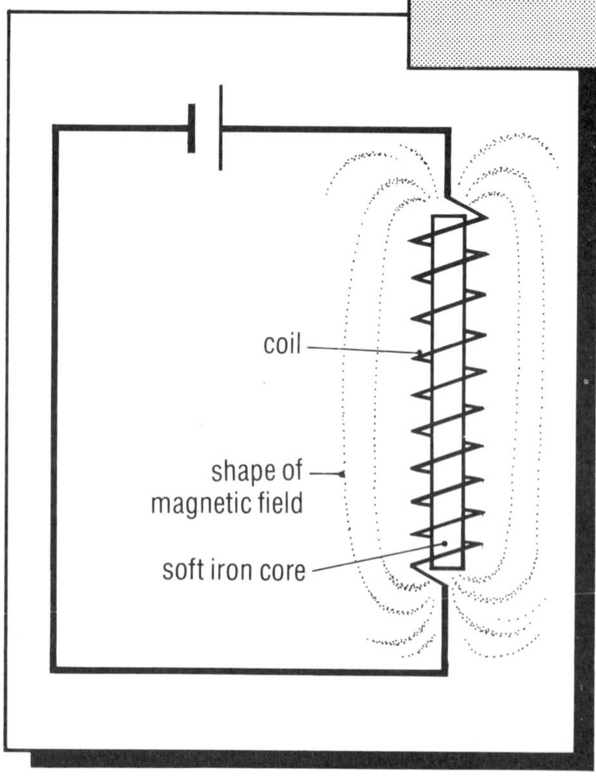

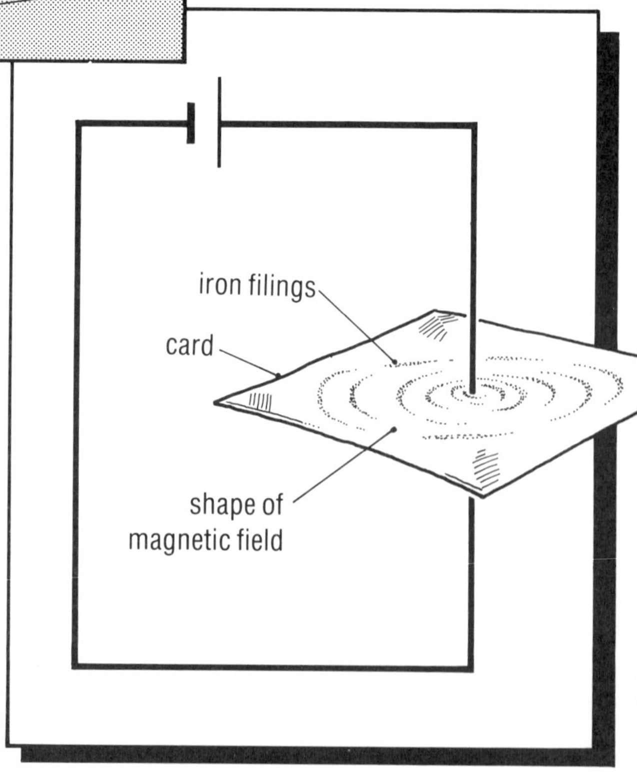

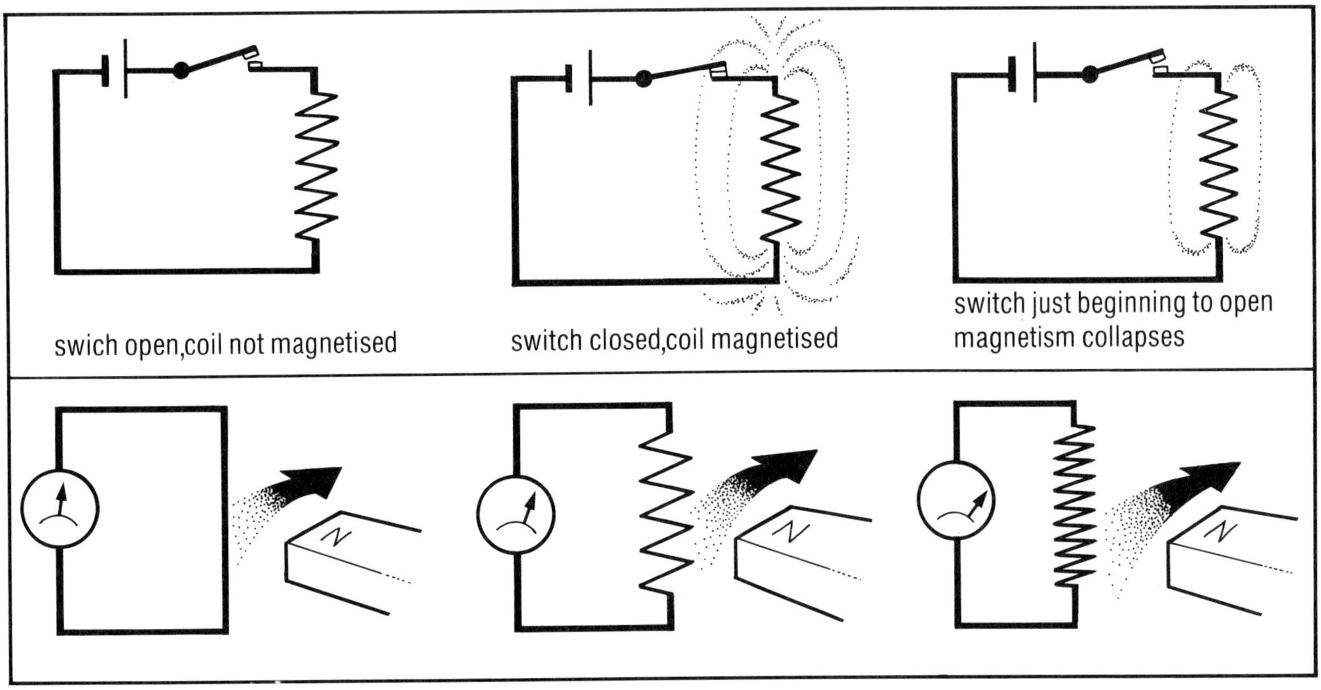

The low-tension (LT) circuit

When the contact breaker points are closed the low-tension current of 12 volts flows and magnetises the primary coil. When the points open the magnetism collapses.

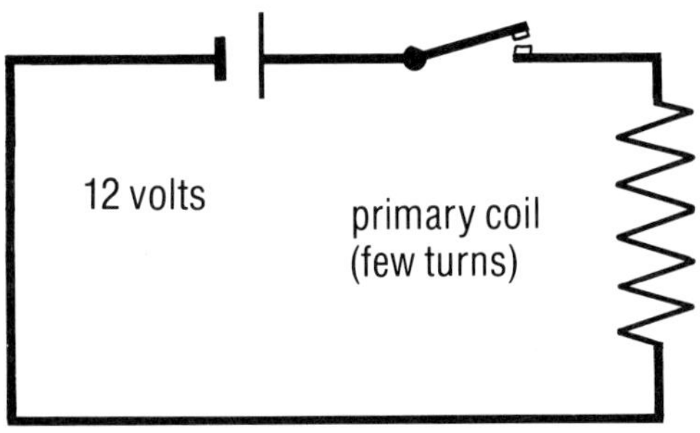

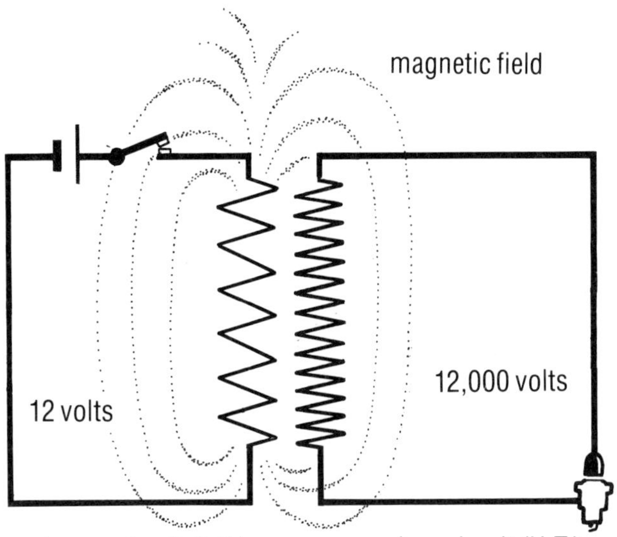

The high-tension (HT) circuit

As the magnetic field collapses it passes through the windings of the secondary coil and a high-tension current of 12 000 volts is induced.

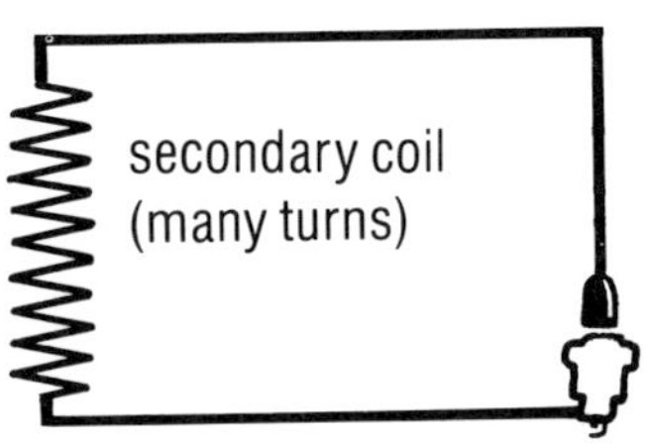

QUESTIONS

1. Draw the circuit diagrams for the LT and HT circuits.
2. Explain how the ignition coil operates.

14. The distributor

In a single cylinder engine no distributor is needed. Multi-cylinder engines have to ensure not only that the contact breaker points open and close at the correct time but also that the HT current goes to the correct spark plug.

The distributor does two jobs:

1 It opens and closes the contact breaker points
2 It distributes the HT current to the spark plugs.

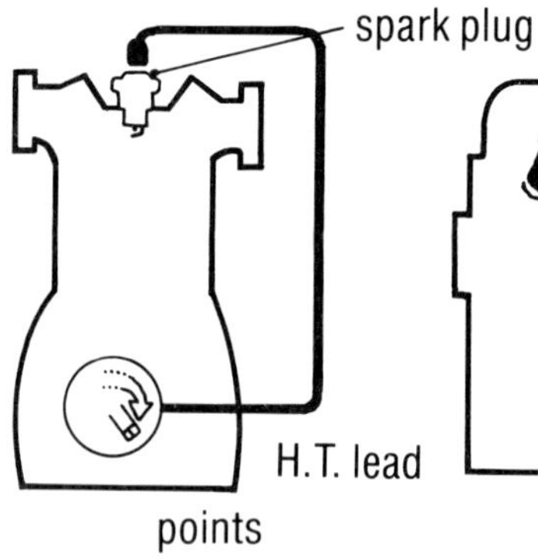

SINGLE CYLINDER ENGINE

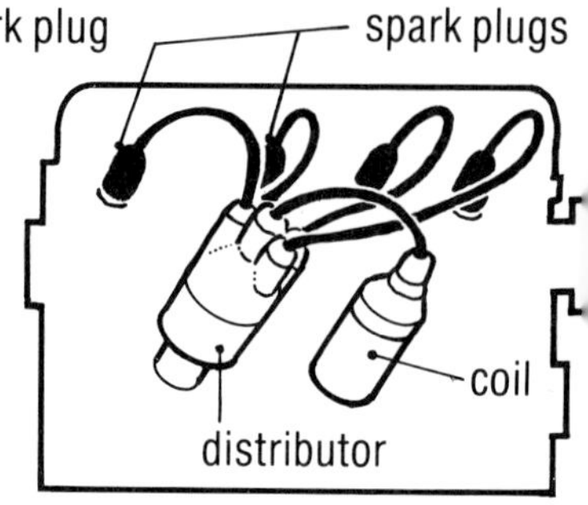

MULTI-CYLINDER ENGINE

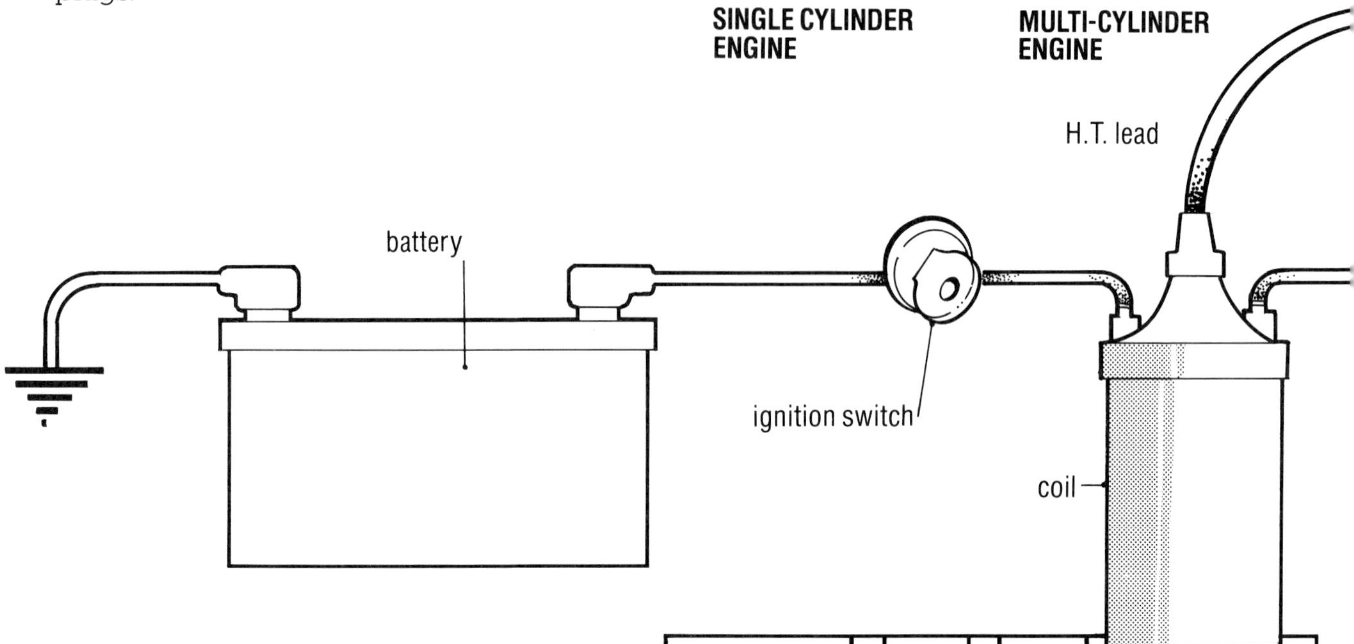

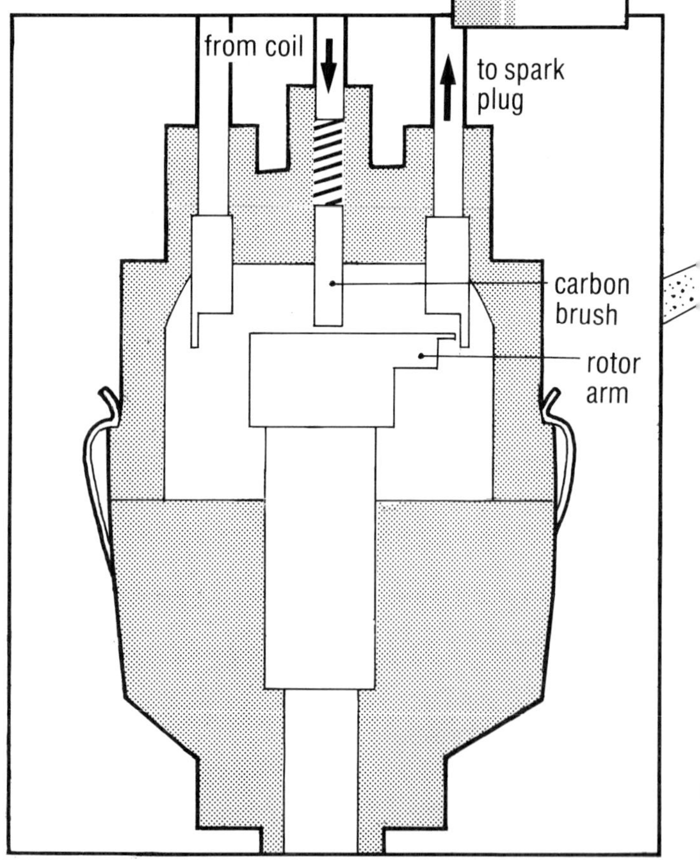

1 The distributor in the LT circuit

The spindle in the distributor is normally driven by a gear on the camshaft and acts as a cam to open and close the contact breaker points (four times in one revolution on a four-cylinder engine, six times on a six-cylinder). The distributor makes sure that the LT current is switched off as the piston approaches t.d.c. on the compression stroke.

2 The distributor in the HT circuit

The distributor distributes HT current to each spark plug in turn. The HT current must go to the spark plug in the cylinder where the piston is approaching t.d.c. on the compression stroke. To keep the valves and distributor in time with one another the distributor is driven by a gear on the camshaft. This gear turns the spindle, on which is fitted a rotor arm. HT current arrives through the top of the distributor cap, passes through a carbon brush and along the rotor arm to the HT lead to the spark plug.

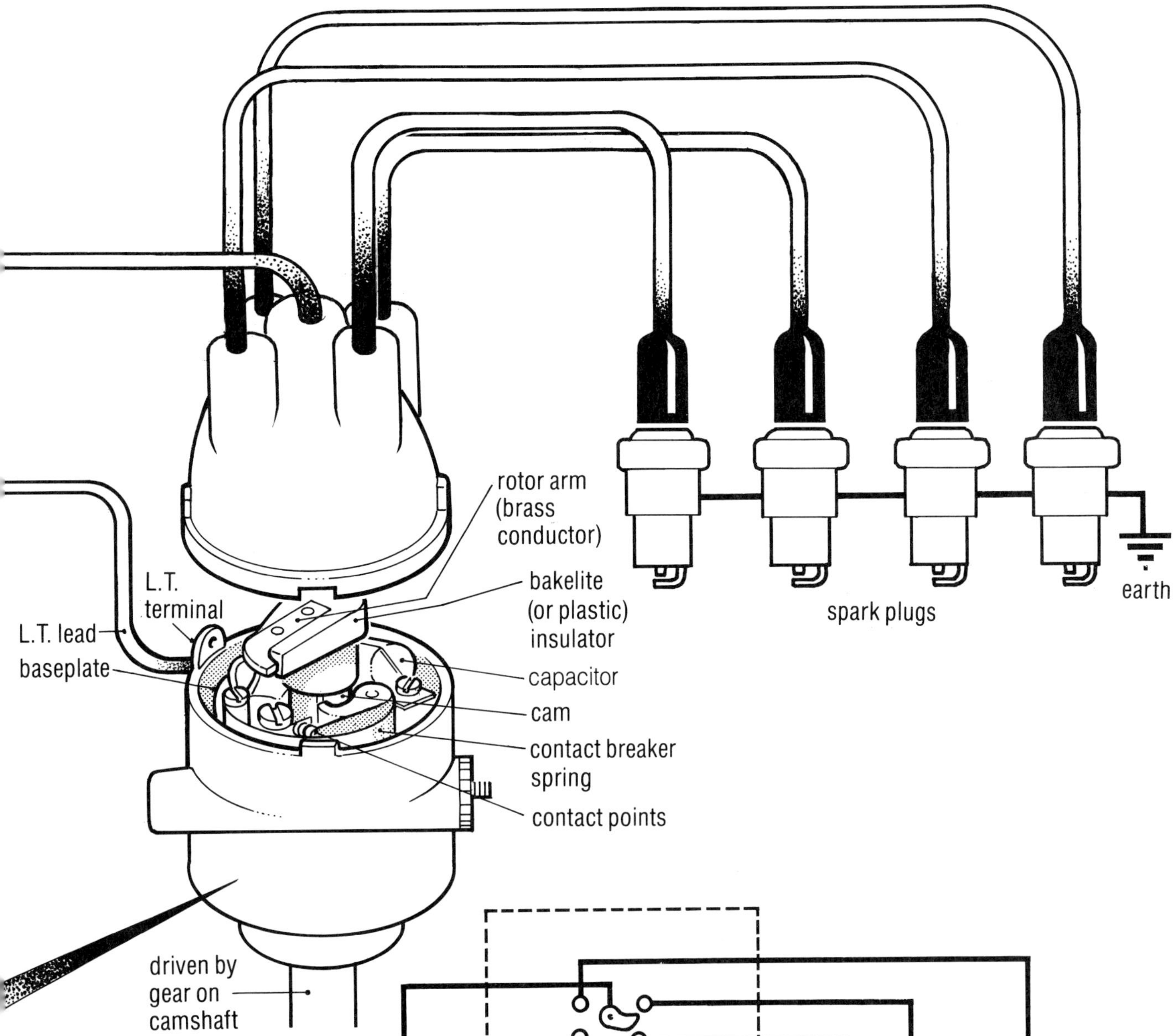

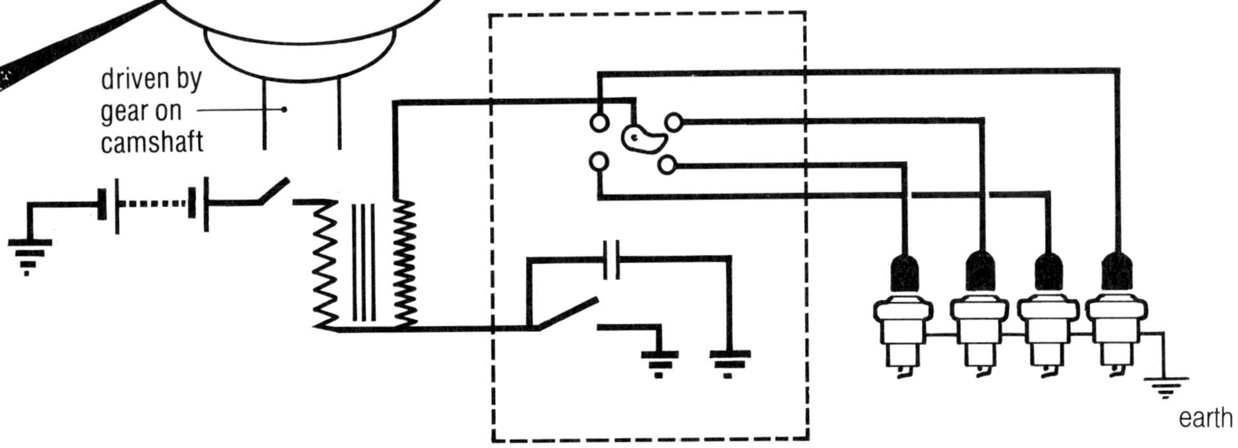

QUESTIONS

1 Make neat, labelled diagrams of the LT and HT parts of the distributor.

2 Make a neat sketch of the complete ignition system and explain what the distributor has to do.

Practical: Remove distributor cap, check for signs of 'tracking' and segment wear. Check condition of HT leads.

15. Contact breaker points

The contact breaker points are a switch which controls the LT current.

Contact breaker points closed: LT current flows into the primary coil and creates a magnetic field.
Contact breaker points open: LT current switched off. Magnetic field collapses and induces a short burst of HT current.

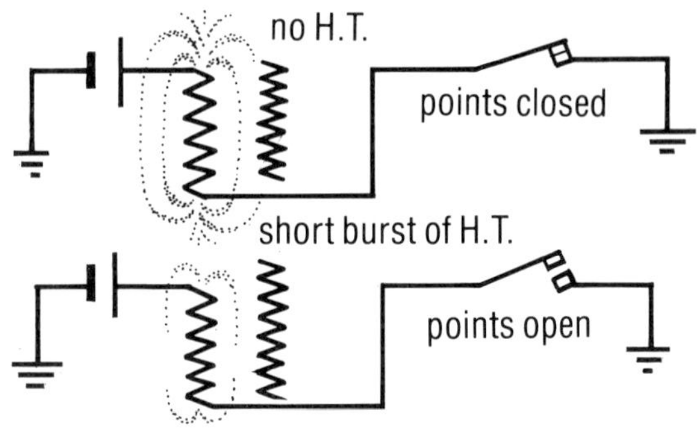

The contact breaker points are opened and closed by a cam in the distributor. They must be closed for long enough for the coil to build up enough current to induce the HT current. The amount of time for which they are closed is set by adjusting the gap between the fixed contact and the moving contact. The contact breaker gap is set using either a *dwell meter* or *feeler gauges*.

A dwell meter measures the amount of movement of the cam for which the points are closed (the dwell angle).

Feeler gauges are strips of steel of various thicknesses. The points are adjusted until the appropriate feeler gauge (usually 0.4–0.6 mm) will just slide between the fixed and moving contacts.

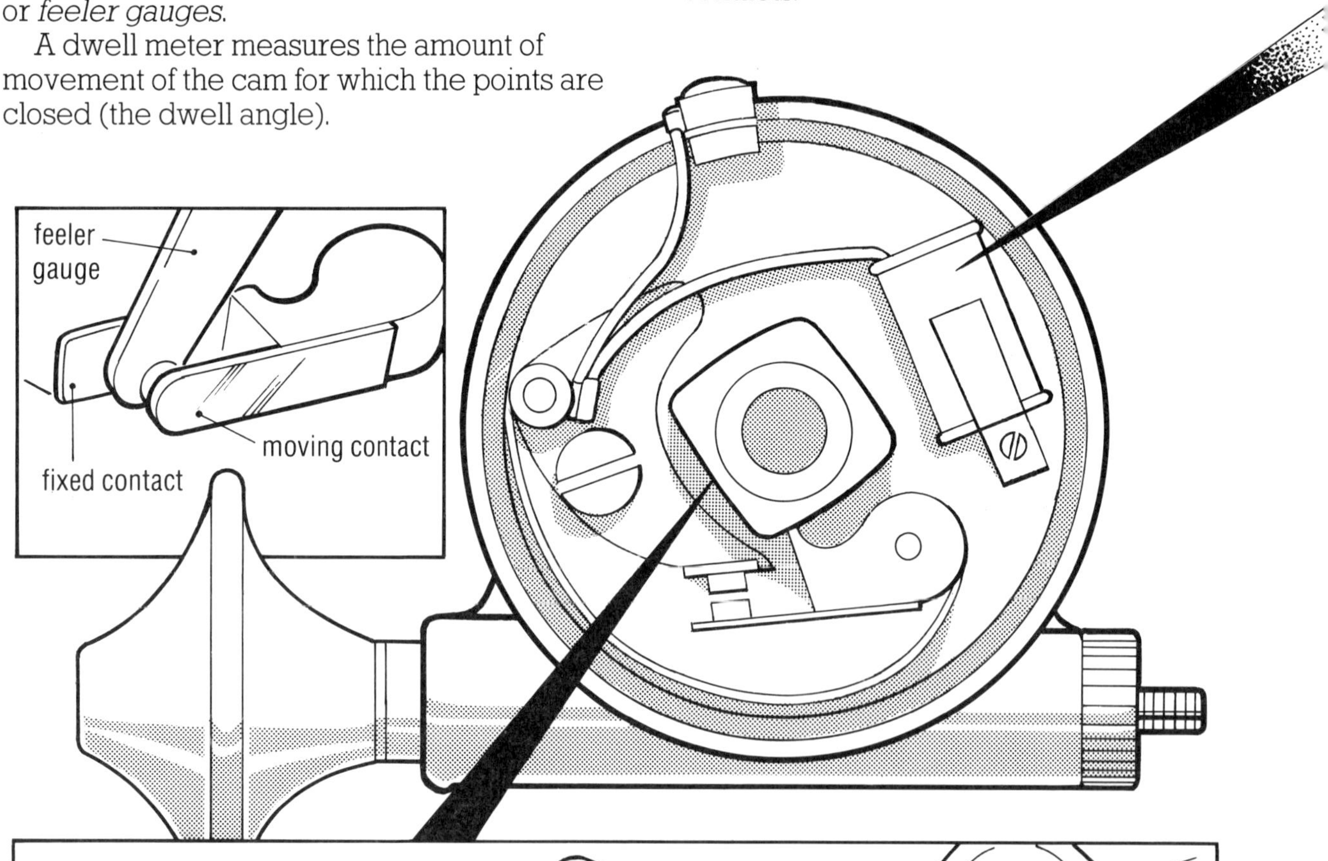

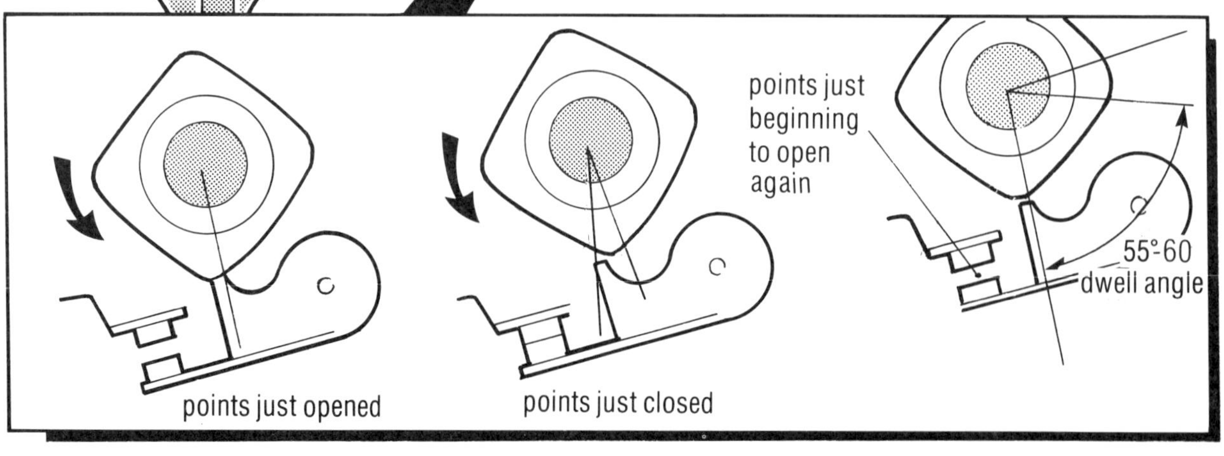

Condenser (capacitor)

As the points begin to open there is a tendency for them to spark, causing burning and wear. The condenser (sometimes called capacitor) prevents this. The condenser is a store for electricity. As soon as the points begin to open the electricity goes into the condenser, as this is easier than sparking across the air gap. As the points begin to close the stored electricity flows from the condenser and helps to start the current flowing through the circuit again. Burned or pitted points are a sign of a faulty condenser.

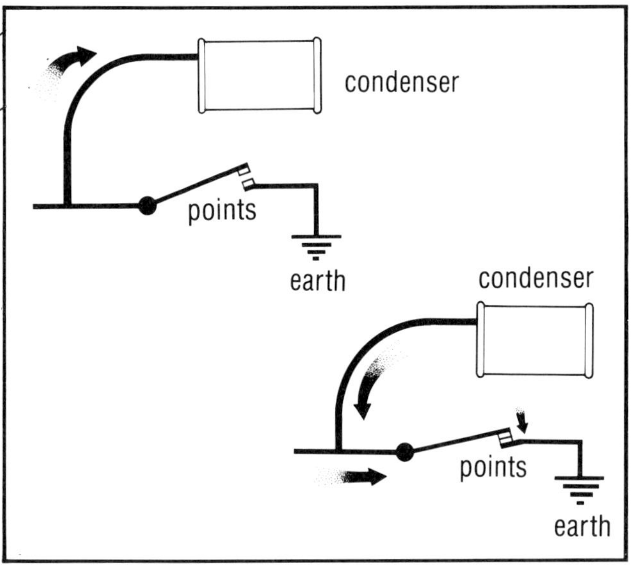

Electronic ignition

Contact breaker points tend to wear, the gap alters and performance is affected. Electronic switches are better because they do not wear. Electronic ignition gives more consistent performance, reliability, better starting and better high-speed performance because they do not 'bounce'.

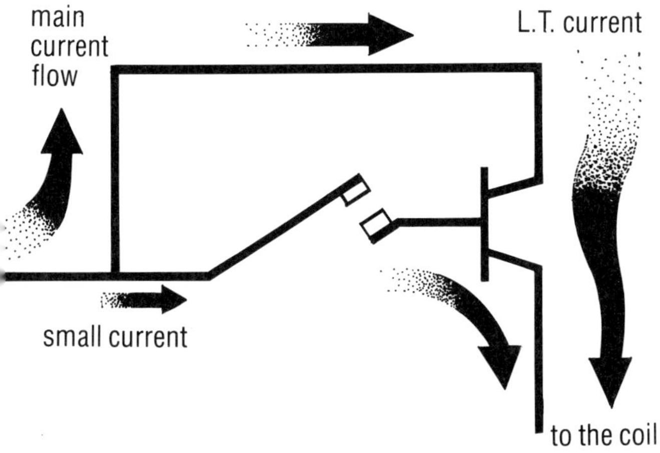

TRANSISTORISED ELECTRONIC IGNITION
A transistor is an electronic switch. In transistorised ignition systems the contact breaker points are used to control a small current through a transistor. The transistor controls the LT current to the coil. A small current flowing through the transistor from the points switches on the LT current through the transistor to the coil. The points control a very small current and so do not wear.

CONTACTLESS ELECTRONIC IGNITION
The contact breaker points are replaced by an optical or magnetic switch.

1 *Optical:* a light emitting diode sends a beam of light to a photo-transistor. When the light beam is interrupted by a tab on a rotating disc the transistor automatically switches the LT current to the coil.

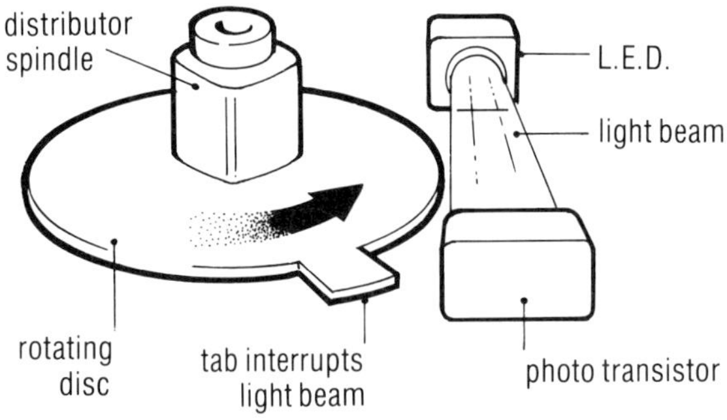

2 *Magnetic:* a magnet sends a short pulse of magnetism to a control unit with transistorised switches to control the LT current to the coil.

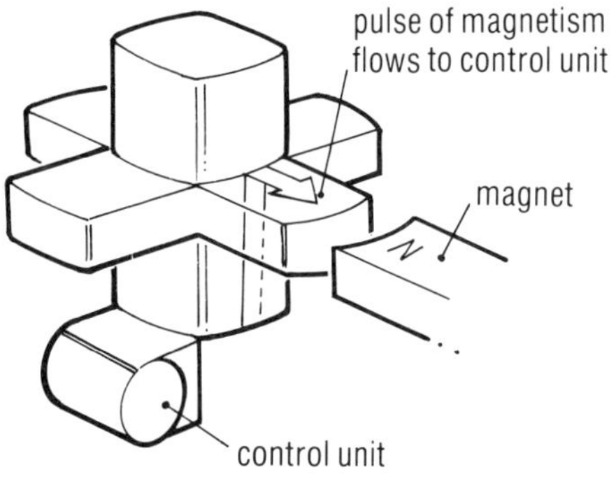

QUESTIONS

1 Make diagrams and notes to explain what happens as the contact breaker points open and close.
2 What are the advantages of electronic ignition?

Practical: Remove, examine, replace and set gap on contact breaker points.

16. Timing

The spark has to occur at exactly the right time as the piston nears the end of the compression stroke. The spark is timed to occur as the piston approaches top dead centre. This gives time for the flame to spread through the mixture and develop maximum power as the piston reaches t.d.c.

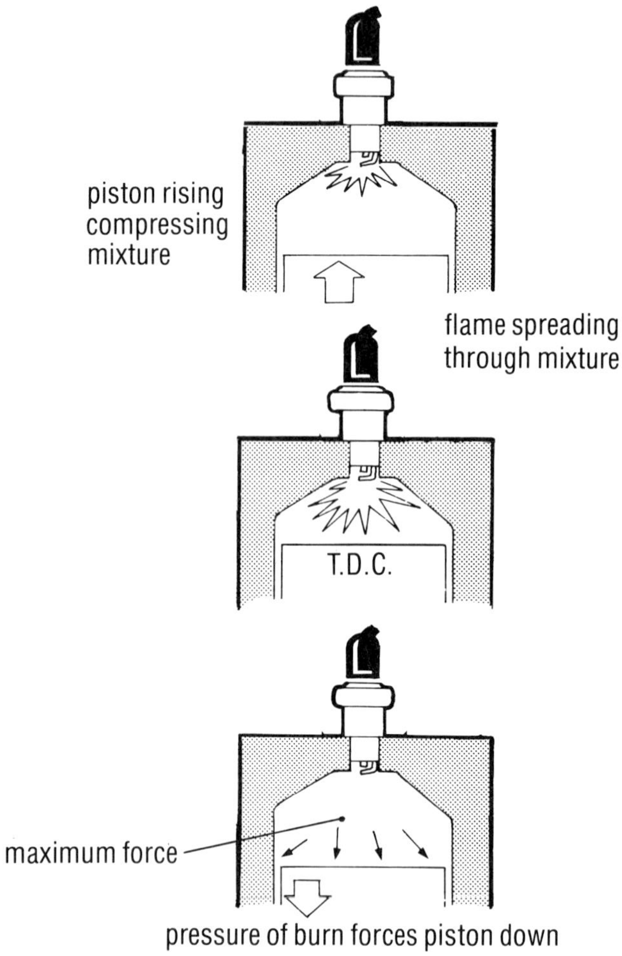

Advanced timing: The spark occurs early, the mixture burns and pushes against the rising piston, giving less power.

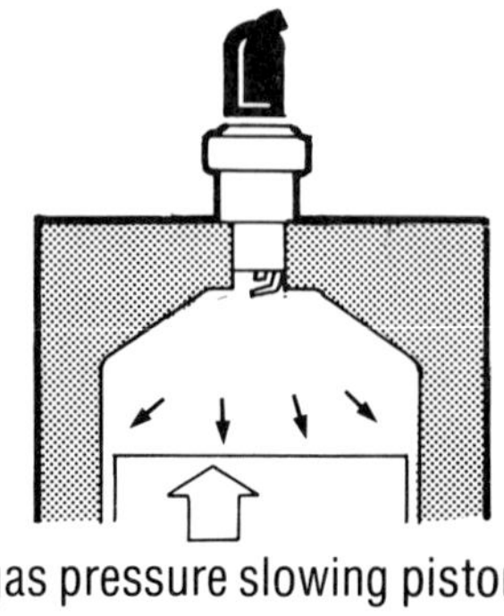

Retarded timing: The spark occurs late, the mixture burns as the piston is moving away and there is less power.

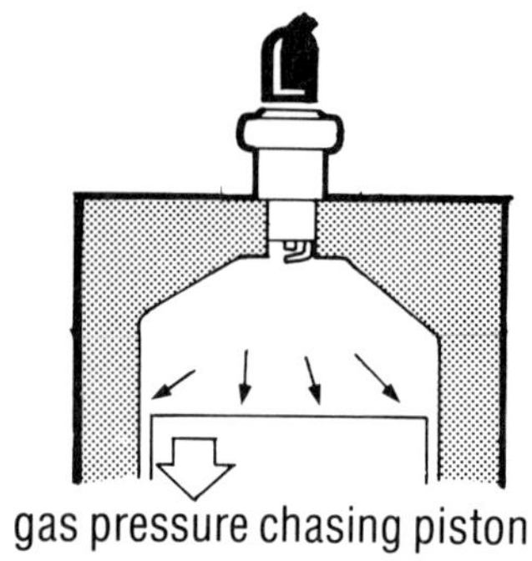

A *stroboscope* is used to set ignition timing. A signal from the spark plug lead turns the stroboscope light on for a split second every time the plug fires. This light is directed at a timing mark on the crankshaft pulley and a pointer. If the timing is correct the mark and pointer will be lined up as the stroboscope light is triggered. If the two do not line up the distributor must be rotated slightly.

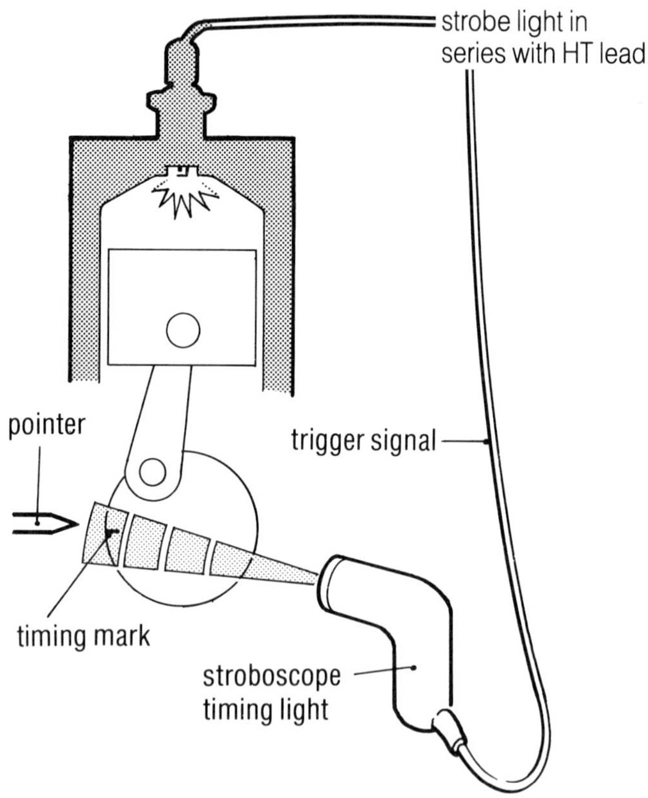

Automatic centrifugal advance

The faster an engine turns the earlier in the compression stroke it needs to fire to allow for complete burning of the mixture. Two bob weights are fixed to the distributor spindle and as it rotates faster they are thrown outwards by centrifugal force. This movement causes a cam to move which advances the ignition timing automatically. As the engine slows down the weights are pulled back in by springs and the ignition is retarded again.

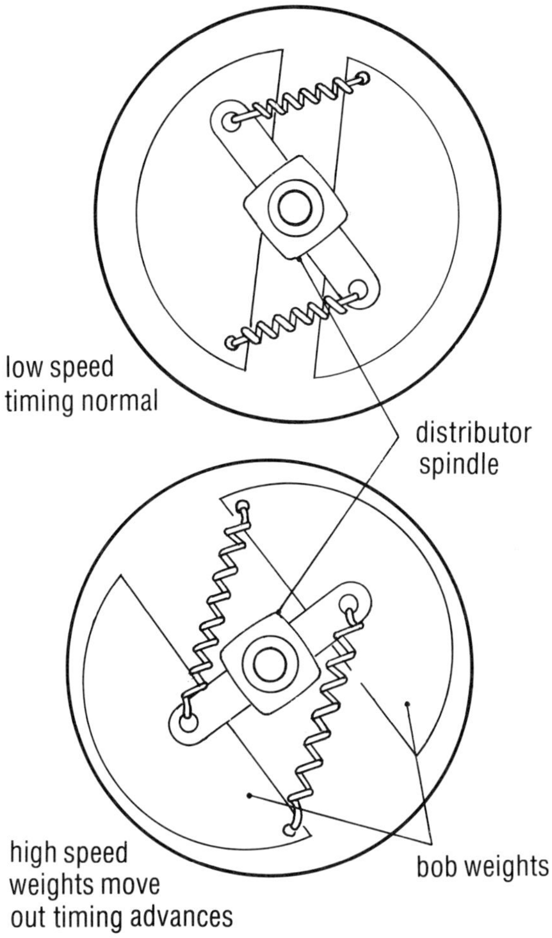

Automatic vacuum advance

The time taken for the mixture to burn depends on how rich or lean the mixture is. A tube connects the inlet manifold to a diaphragm linked to the distributor base plate. As the mixture becomes leaner and more is drawn in, the greater is the suction at the inlet manifold and the greater the movement of the distributor base plate.

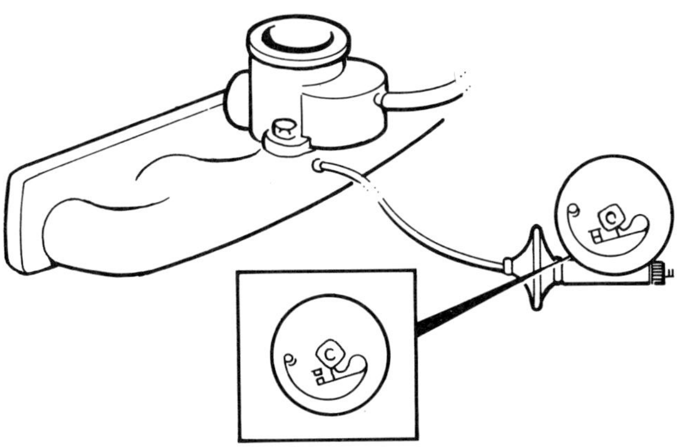

Firing order

The cylinders are numbered, with number 1 cylinder being the one nearest the front of the car or at the end of the engine with the water pump or fan. If the firing order of the engine were 1, 2, 3, 4 the crankshaft would be put under a lot of pressure. A more even load on the crankshaft results from the firing orders:

1, 2, 4, 3 or 1, 3, 4, 2

The chart shows what happens inside each cylinder.

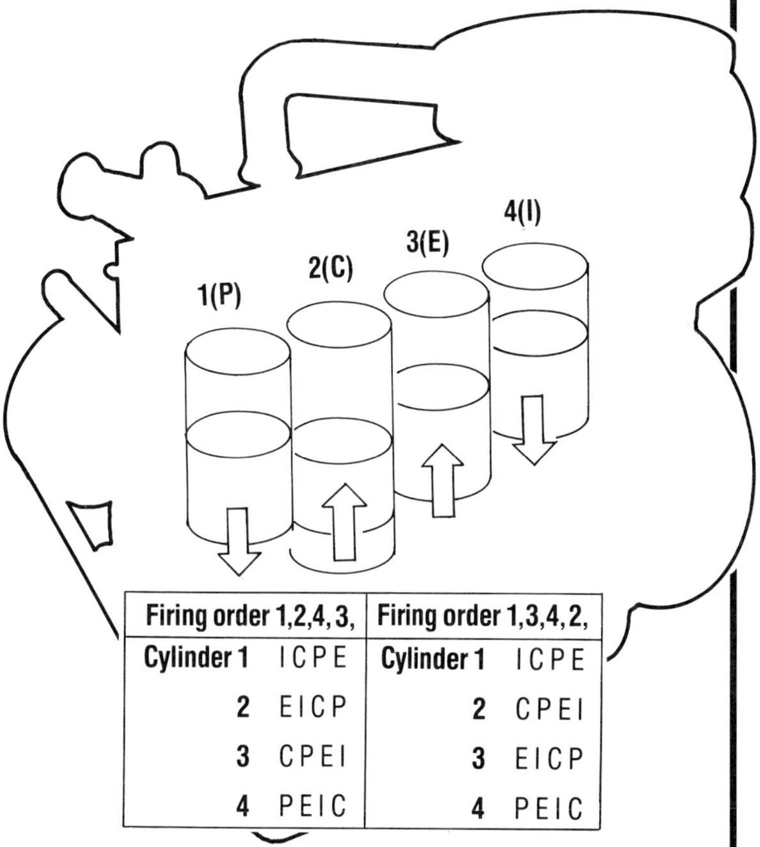

Firing order 1,2,4,3,		Firing order 1,3,4,2,	
Cylinder 1	I C P E	Cylinder 1	I C P E
2	E I C P	2	C P E I
3	C P E I	3	E I C P
4	P E I C	4	P E I C

QUESTIONS

1. Write a short note to explain what is meant by advanced timing and retarded timing. Use diagrams to help and mention both automatic devices.
2. Explain why the firing order is not 1, 2, 3, 4 and what happens in the other cylinders when number one cylinder is on its power stroke in an engine with a firing order of 1, 2, 4, 3.

Practical: Remove, examine and replace timing chain. Set static timing. Set timing using a strobe lamp.

EXAMINATION QUESTIONS

THE IGNITION SYSTEM

1 The purpose of the distributor is to ensure:
a) a spark in each cylinder at the correct time;
b) that the spark in each cylinder is uniform;
c) that the necessary quantities of fuel reach each cylinder at the correct stage of the cycle;
d) that the rotor arm does not overheat. (*NEA*)

2 When servicing a car ignition system a feeler gauge is used to measure the adjustment of the ________ and the ________ (*NEA*)

3 If a sparking plug becomes sooted is the mixture likely to be too rich or too lean?

4 What is the purpose of each of the following:
a) C.B. points
b) distributor cam
c) condensor
d) rotor arm
e) carbon brush

5 Make a neat, labelled diagram of a section through a sparking plug.

6 In a four cylinder engine with a firing order of 1, 3, 4, 2 what stroke will number three cylinder be on when number one is on its power stroke?

7 Make a neat, labelled diagram of an ignition system, including HT circuit, LT circuit, distributor, coil and sparking plugs.

8 Write a detailed explanation of how the ignition coil works and discuss the advantages of electronic ignition.

9 Identify the parts indicated on the diagram of an ignition coil.

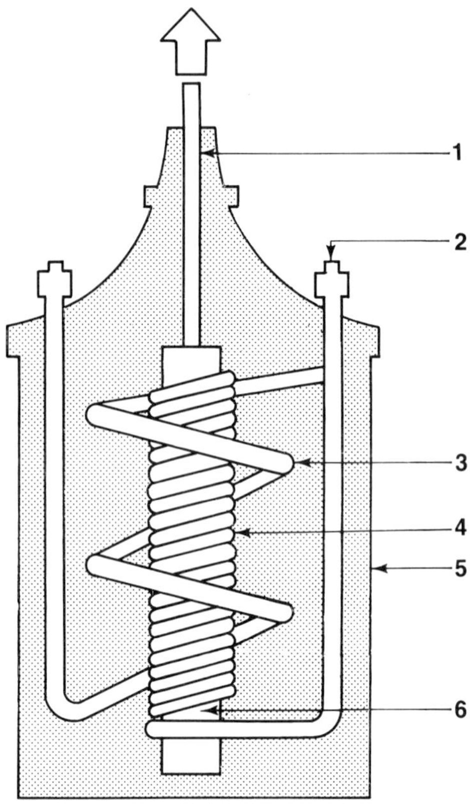

10 a) What is the effect of advanced ignition timing?
b) What is the effect of retarded ignition timing?
c) How does the automatic advance-retard mechanism work?
d) How does the automatic vacuum advance work?

Cooling and lubrication

Cooling using air or water helps maintain an efficient working temperature in the engine and prevents overheating. Lubrication of moving parts with a suitable grade of oil reduces friction and wear as well as assisting cooling.

17. Cooling

Burning petrol inside a cylinder creates a lot of heat. Hot metal expands and if it is not cooled the engine will seize. Two methods of cooling engines are used: air and water. (Water cooling is discussed in the next chapter.)

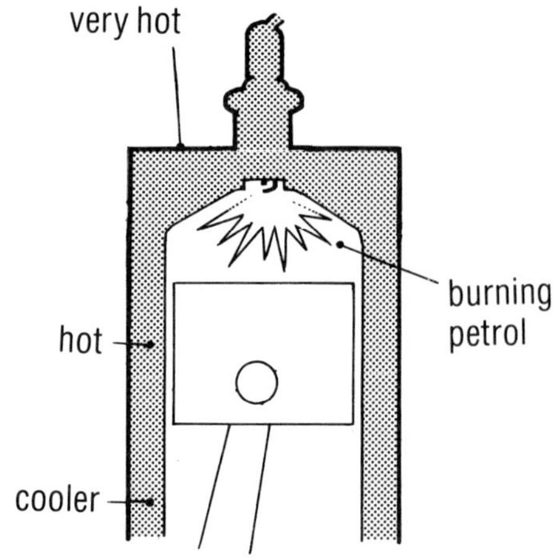

Air cooling

Air cooling is simple and cheap. There are no pumps, radiators, hoses (or leaks!). It works best when the engine is exposed to the air and the vehicle is moving. Air cooling is therefore used mainly on motor cycles. (Cars need a fan and ducting. A large fan will use engine power and air cooled engines are noisier than water cooled engines.)

HOW IT WORKS
A large block of metal will cool slowly.

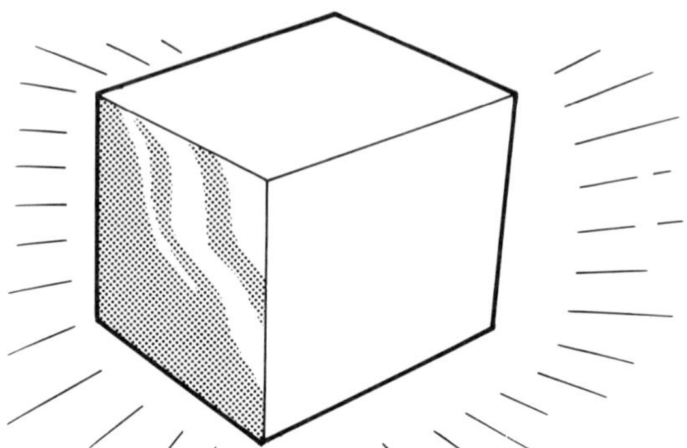

A thin sheet of metal will cool quickly.

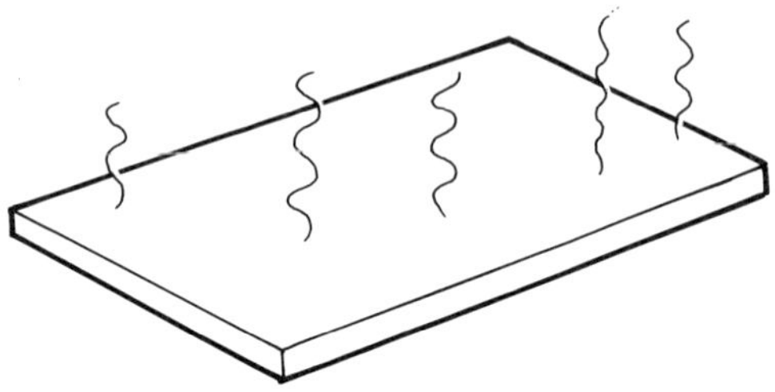

Tapered *fins* help heat flow away from a hot area to a cooler area, and increase the surface area.

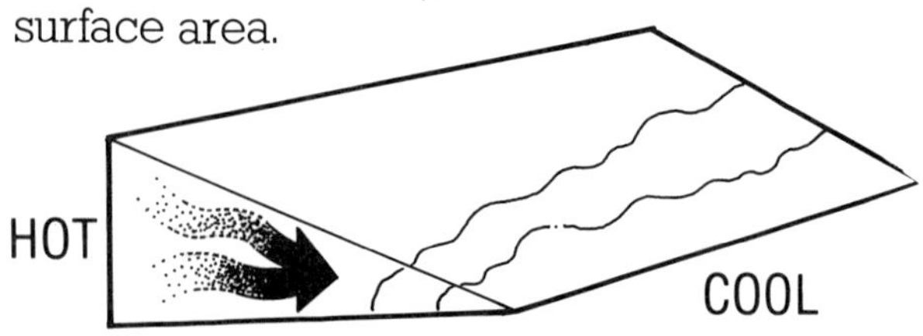

Heat travels from the burning mixture to the cylinder walls, then along the fins, and cool air blows over the fins to cool the engine. The fins at the top of an engine are larger than those at the bottom because there is more heat at the top of the engine.

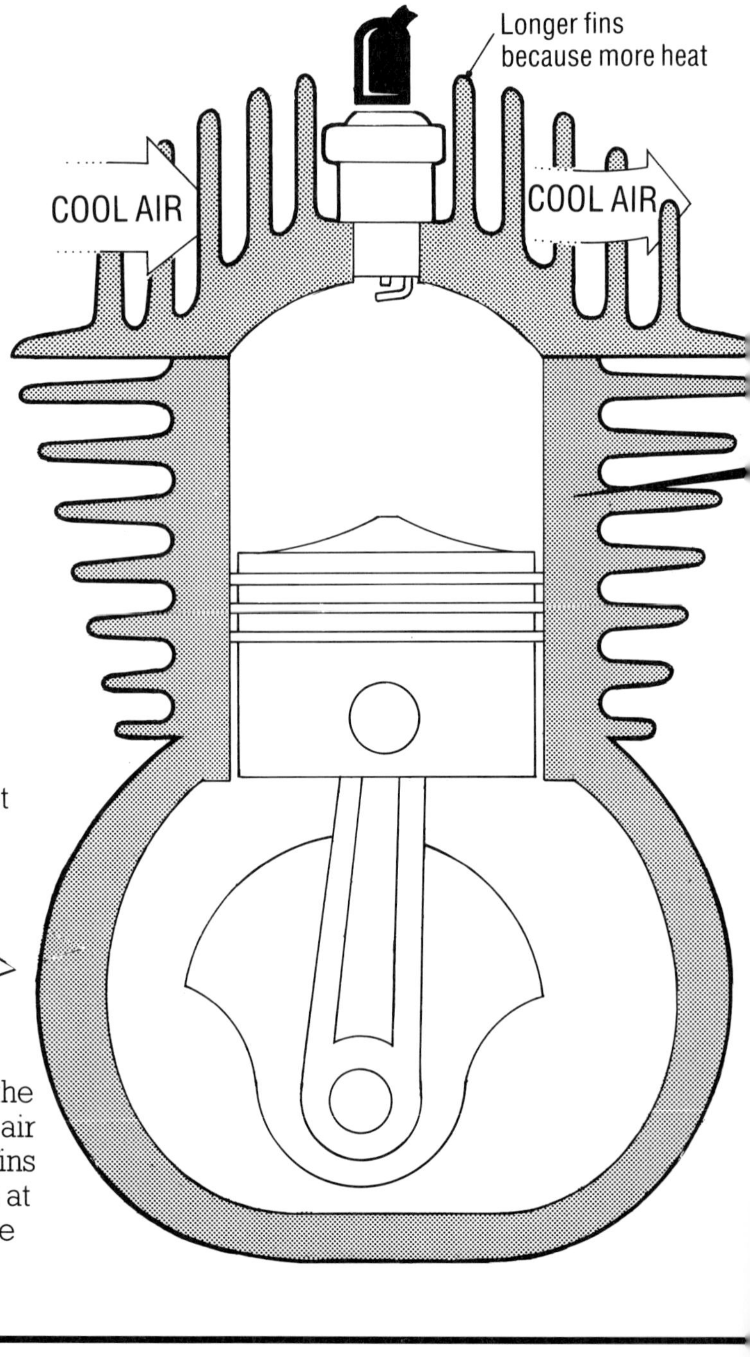

Heat travel

Heat travels in three ways: conduction, convection and radiation.

CONDUCTION

The way heat travels through solids. As molecules in the material are heated they vibrate and bump into other molecules.

CONVECTION

The way heat travels through liquids and gases. Hot molecules are lighter than cold molecules and tend to rise.

RADIATION

The way heat travels through space where there are no molecules.

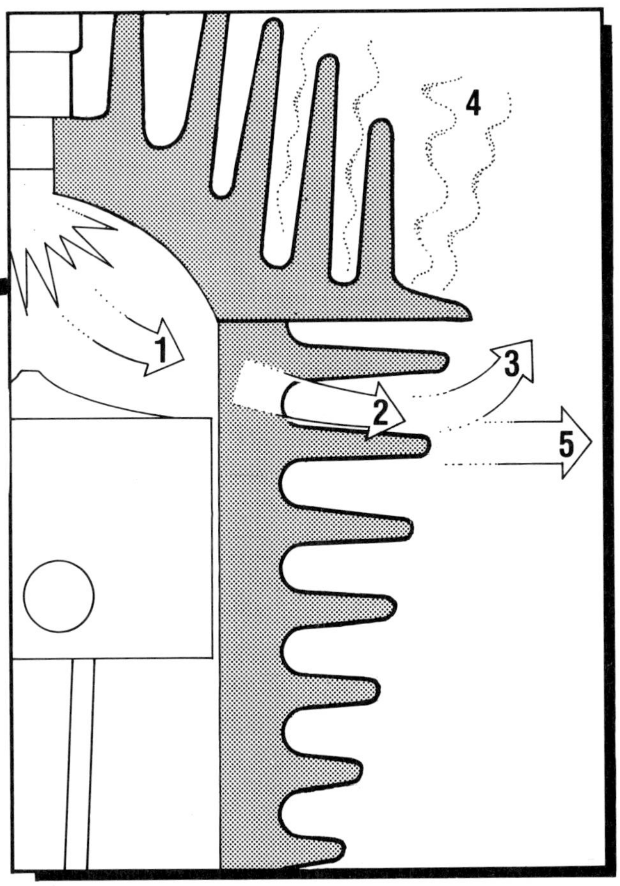

Heat travels by conduction, convection or radiation in different parts of the engine.

1 From combustion to cylinder wall (conduction).
2 From cylinder wall along the cooling fins (conduction).
3 From fins to cool air (conduction).
4 Heat rising from engine (convection).
5 Heat felt at the side of the engine (radiation).

QUESTIONS

1 Write a short explanation of how heat travels by conduction, convection and radiation.

2 Give examples of places in a car where heat travels by conduction, convection and radiation.

18. Water cooling

Cooling an engine by using water passing through passages in the cylinder block (a water jacket) has several advantages:

1 It is easier to keep an even temperature around each cylinder.
2 Surrounding an engine with water reduces noise.
3 The hot water can be used in the car heater.

Heat passes from the cylinder to the water by conduction, the hot water rises by convection, passes through the top hose to the radiator, where it cools before passing back to the engine through the bottom hose. The cooling fan draws air through the radiator and passes it over the outside of the cylinder block to provide additional cooling. The fan also drives the water pump, which ensures an even circulation of water.

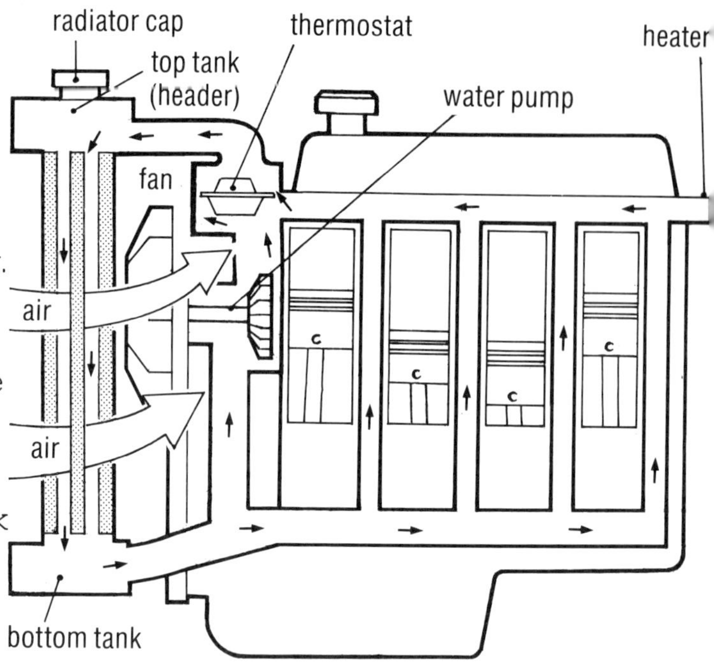

The radiator

The radiator consists of a top (header) tank, thin tubes of brass, fins (to increase the surface area for cooling) and a bottom tank. The radiator is connected to the cylinder block by flexible rubber hoses which allow for engine movement. As the hot water passes from the header tank through the tubes to the bottom tank it is cooled by air passing over the tubes and connecting fins.

Pressurised cooling system

Engines run best at a water temperature of 85 °C–110 °C. Water boils at 100 °C. To allow engines to run at higher temperatures the water cooling system is pressurised. A pressure cap is fitted to the radiator so that before the water can boil it must raise enough pressure to lift a valve against a spring pressure. This means that the water will not boil until about 110 °C.

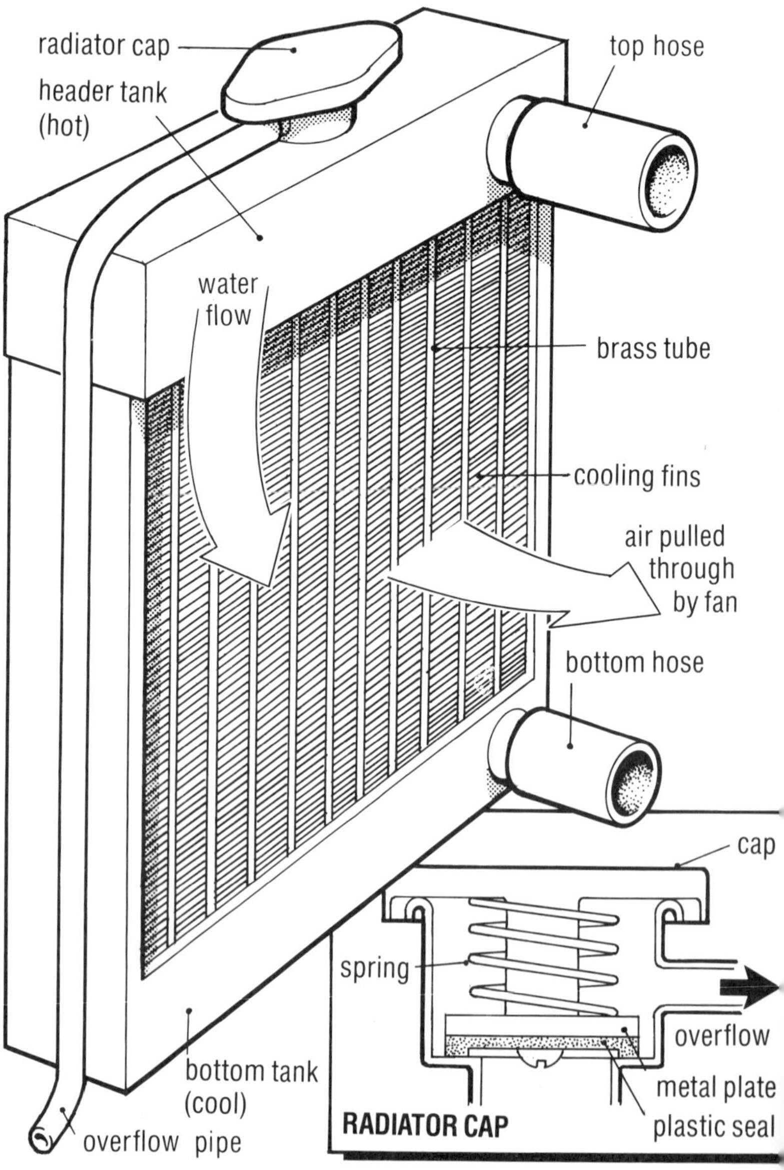

RADIATOR CAP

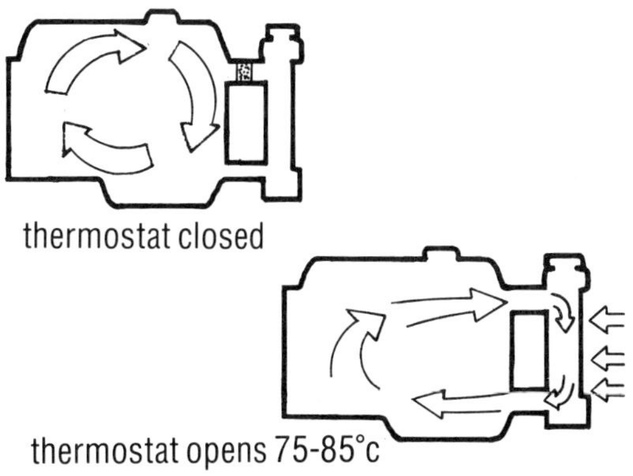

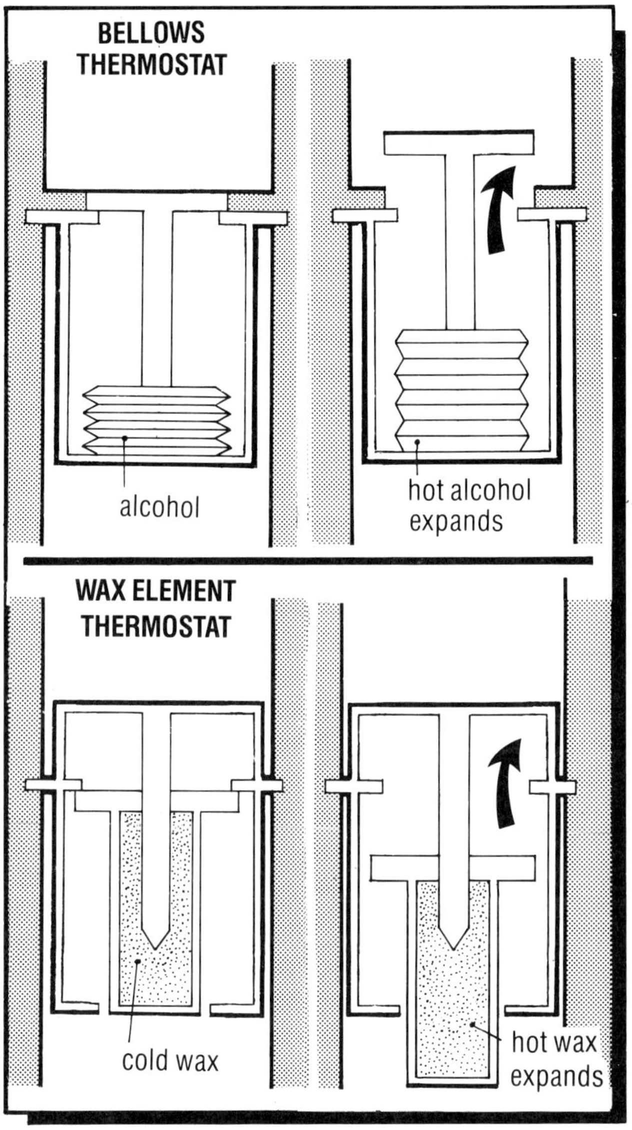

The water will only stay liquid at these higher temperatures as long as the sytem is sealed and pressurised. If the radiator cap is removed while the water is still hot the reduced pressure will allow the water to boil and produce a cloud of scalding steam. [*NB Do not remove the radiator pressure cap until the engine is cool!*]

Thermostat

A cold engine is not very efficient. An automatic valve called a thermostat helps the engine reach its best operating temperature quickly. When closed it stops water from passing into the radiator and ensures a rapid warm up. When the water heats up to, say, 85 °C the thermostat opens and water passes to the radiator for cooling. If the temperature falls to below, say, 80 °C the thermostat will close again.

There are two types of thermostat:

1 BELLOWS TYPE
Alcohol expands in a metal bellows and opens the valve; as it cools and contracts the bellows close.

2 WAX ELEMENT TYPE
A wax pellet expands and contracts with temperature changes and opens or closes the valve.

Anti-freeze

When water freezes it expands and forms ice, which can burst radiators and cylinder blocks. Anti-freeze lowers the freezing point of water and ensures that any ice that does form is soft slush. A 25%–30% solution of ethylene glycol is usually recommended for British winters. Of course, if you do your motoring at the North Pole you may need 100% anti-freeze!

Disadvantages of water-cooling

The disadvantages of water cooling are that it adds to the cost and weight, is a source of leaks and needs attention (topping up in summer and anti-freeze in winter).

QUESTIONS

1 List the advantages and disadvantages of water cooling and air cooling.
2 Make a neat, labelled diagram of the water cooling system and explain why it is pressurised.

Practical: Drain water cooling system, remove top hose and thermostat cover. Remove, test and replace thermostat. Refit hose and refill system. Run engine and check for leaks.

19. Lubrication

Properties of oil

Oil does four things:

1 Reduces friction
2 Reduces wear
3 Carries away metal and carbon particles
4 Cools the surfaces.

When metal surfaces are moved in contact with one another they wear vary rapidly and get hot. A very thin (0.05 mm) layer of oil is enough to hold the surface apart leading to less friction (which saves engine power) and less wear.

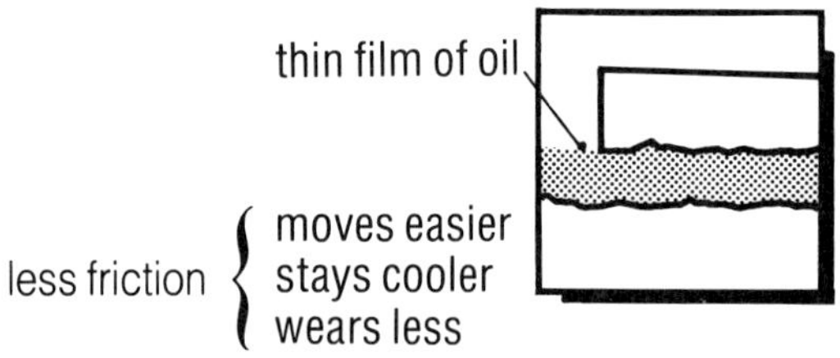

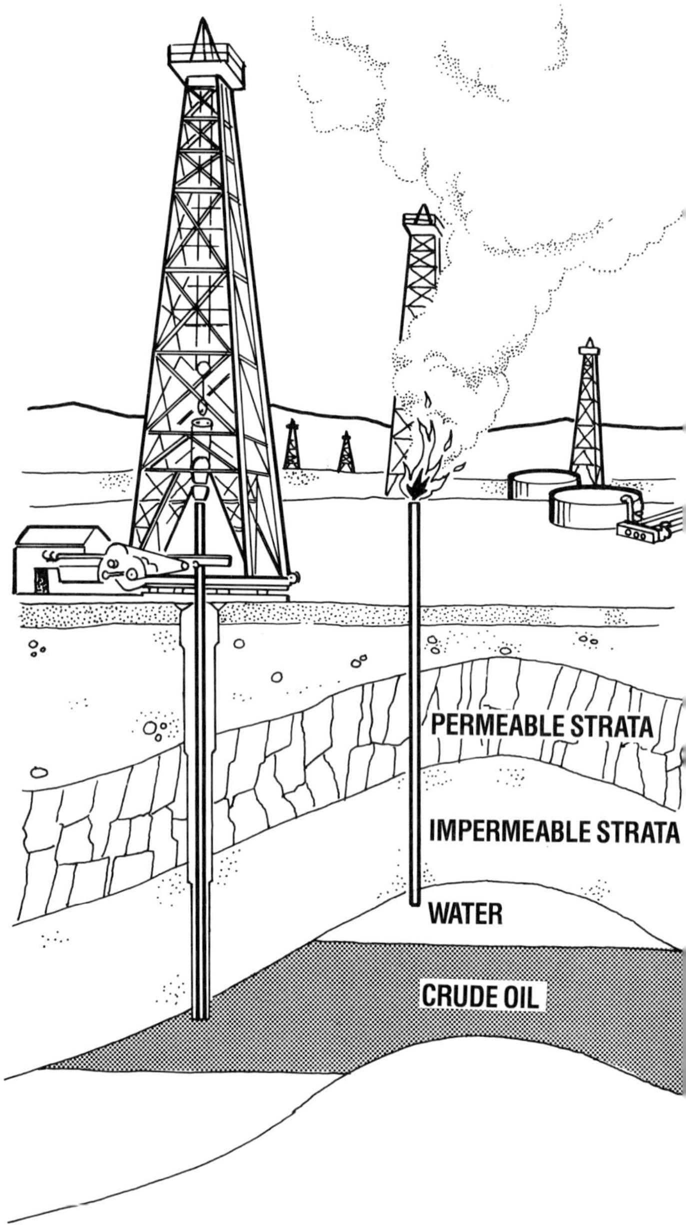

The oil that is used in cars is a *mineral* oil. Crude oil, obtained by drilling into the earth, is refined and various additives are put in to make a good quality lubricating oil.

Viscosity

Oils are graded by their thickness. The Society of Automotive Engineers designed a test which measures the thickness of oil and gives it a number. SAE 20 is thin oil, SAE 30, 40 and 50 are thicker. The technical term for 'thickness of oil' is *viscosity*. The SAE test measures viscosity at 99 °C; the higher the SAE number the higher the viscosity and the thicker the oil. Thin oils are used in colder conditions and are tested at −18 °C. SAE 5W, 10W and 20W are thin oils, their viscosity is low (the W stands for winter).

If a ball bearing is dropped into a high viscosity (thick) oil it will sink to the bottom of the container slower than through a low viscosity (thin) oil. Thin oil does not stick to the steel and will be less effective as a lubricant. Thick oil will tend to slow down any movement and may not be easy to pump to small areas.

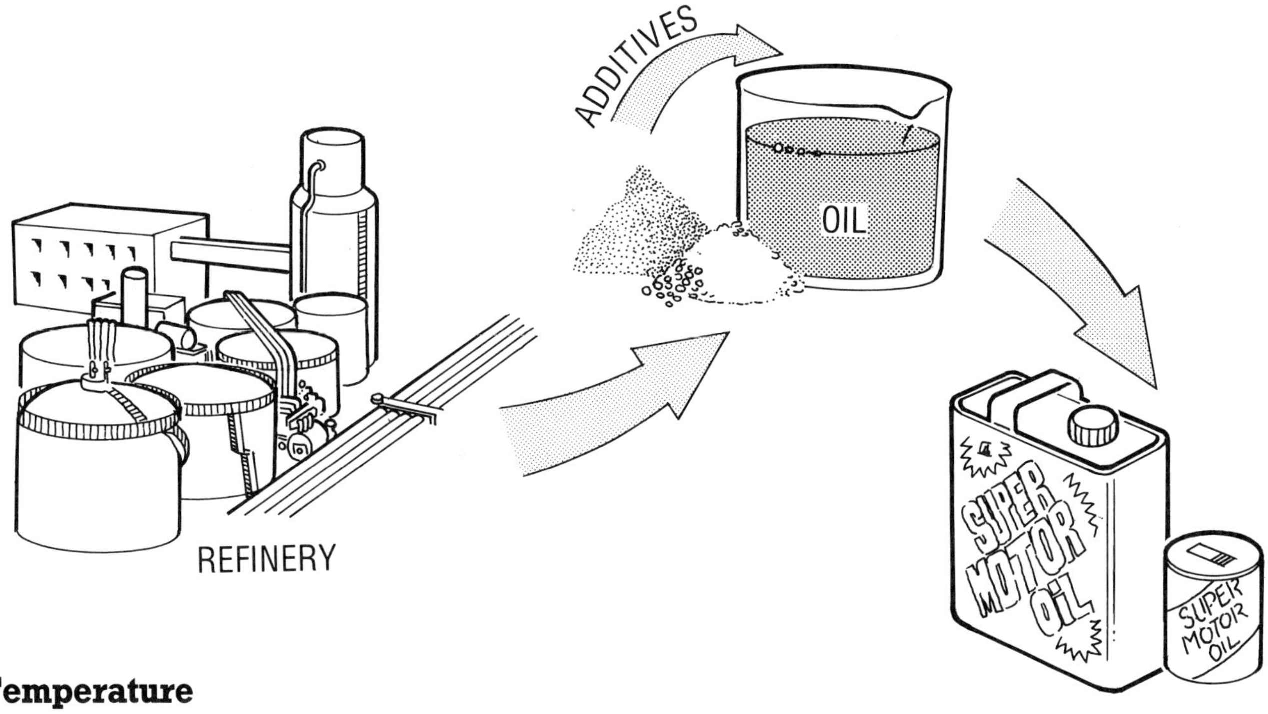

Temperature

Cold oil is thicker than hot oil. The problem is that engines start cold and get hotter as they are run. Motor oils have to be thin when cold to pump quickly to all parts of the engine but thick when hot to keep the metal surfaces apart. Multigrade oils are made by blending oils and using additives. They are thin when cold and get thicker as the temperature increases (the viscosity increases as temperature increases). The SAE numbers for multigrade oils show the viscosity both at −18 °C and at 99 °C, e.g. SAE 20W–50, SAE 10W–50.

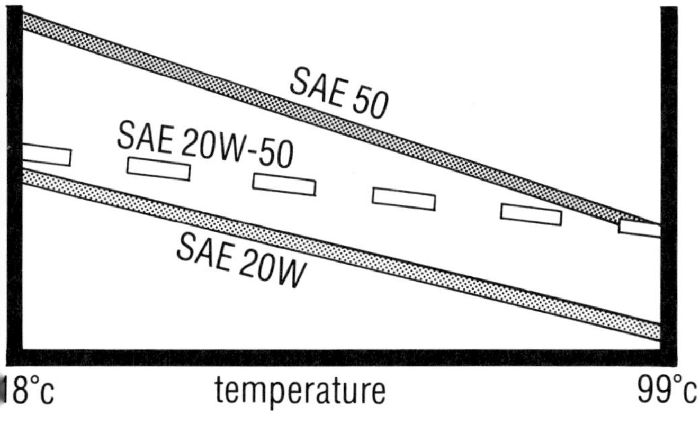

Additives

1 *Viscosity additives* reduce the tendency to be thick when cold and thin when hot.
2 *Dispersants/detergents* prevent soot, dust and carbon from forming sludgy deposits.
3 *Anti-oxidant* prevent oxygen and oil combining to form a thick laquer.
4 *Load carrying additives* improve bearing properties.

Other additives reduce foaming, corrosion and build-up of acids. Because the detergent in the oil gets used up, oil changes are necessary periodically.

Special oils

Extreme pressure (EP) *oils* are used in gearboxes and final drive units to cope with the pressure between gears in mesh.

Grease (lubricating oil mixed with soap) is a lubricant which can be packed around bearings to give long-term lubrication.

Hydraulic oils are light oils strengthened with additives which enable them to transmit the forces in brake or clutch systems.

QUESTIONS

1 Make notes on the four things that oil does in an engine.

2 Explain what is meant by viscosity and how this is affected by temperature.

20. Lubrication systems

There are three types of lubrication system:
1 Wet sump
2 Dry sump
3 Petroil.

Wet sump

Used in most cars. Oil is stored in a pressed steel *sump*, a *pump* picks it up and passes it through a *filter* to remove any particles of metal/carbon and then forces it through *oilways* to the bearings. The oil then drains back into the sump where it cools before being used again.

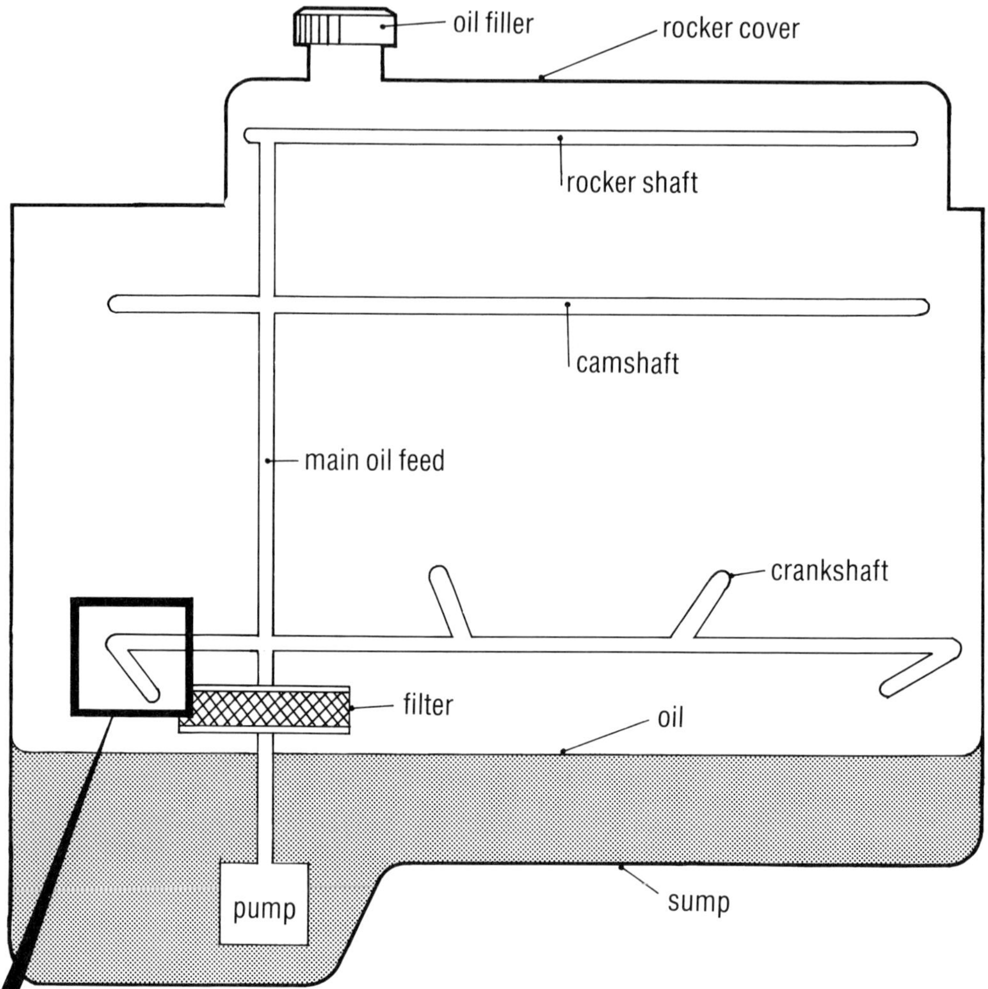

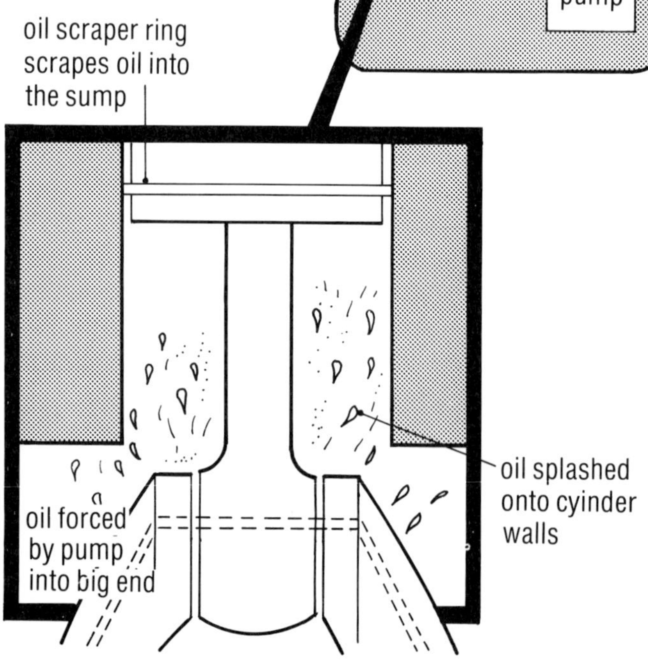

The main oil supply goes to the parts of the engine which take most load: the crankshaft main bearings and big end bearings. The oil goes through passages drilled through the crankshaft. After passing through the bearing, the oil is splashed onto the cylinder walls by the crankshaft. (Crankshaft and big end bearings are *pressure lubricated* the cylinder walls are *splash lubricated*.) The bottom piston ring is an oil scraper ring which prevents the oil reaching the combustion chamber.

From the main pressure supply a feed goes to the camshaft and another to the rocker shaft and timing gear.

Dry sump

Used mainly in motorcycles, this sytem is similar to wet sump except that the oil is stored in a separate tank, pumped around the engine and then returned to this tank.

Petroil

A simple system that needs no pump, filter or oilways, it is used in two-stroke engines, which have no valve gear to be lubricated.

Oil is mixed with petrol (about 20 parts petrol to 1 part oil is a common two-stroke mixture) to form 'petroil'. This is mixed with air by the carburettor to form a sticky mist which is drawn into the crankcase where the drops of oil stick to the metal surfaces, lubricating them. Unused lubricant passes into the cylinder where it is burnt.

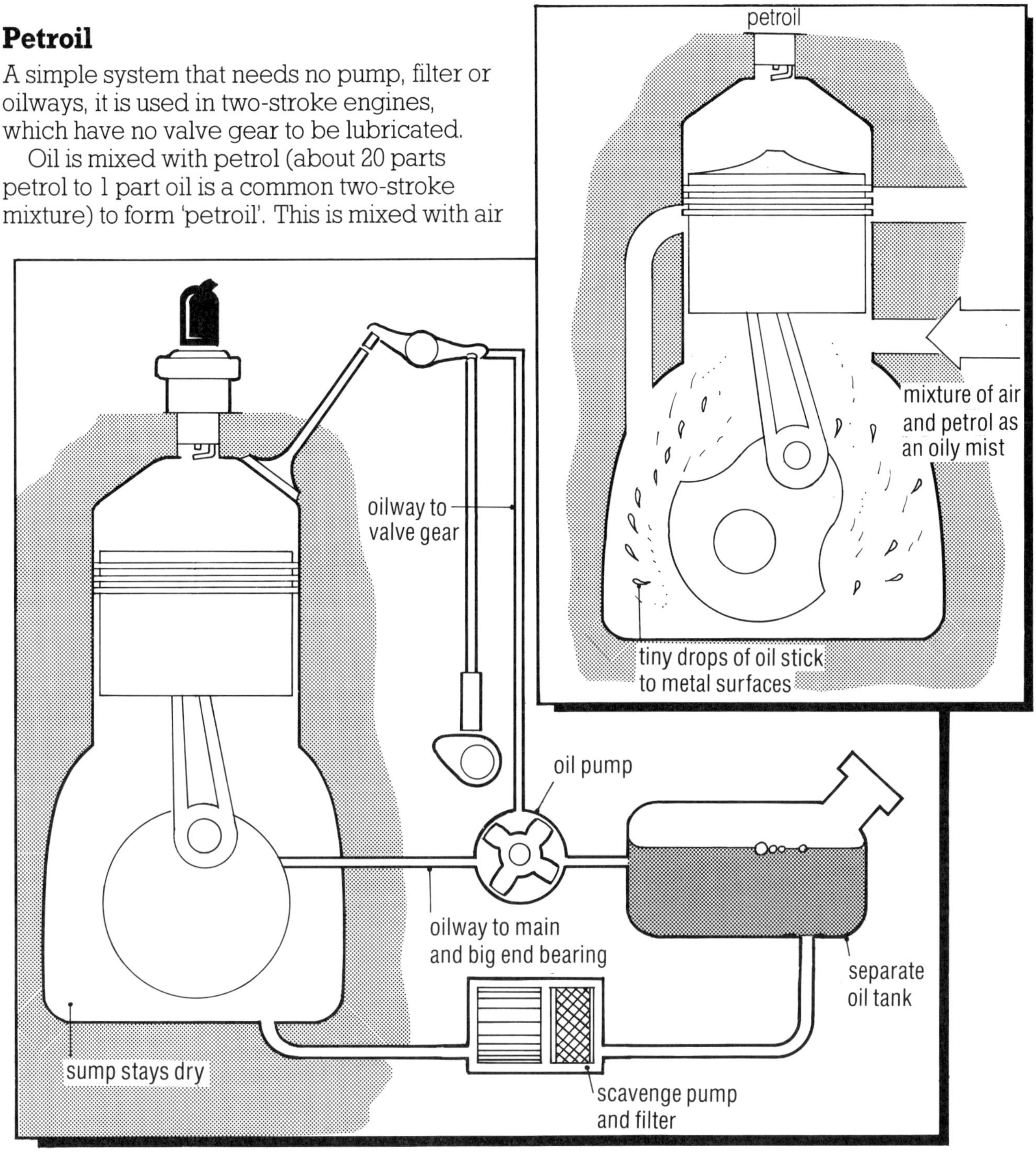

QUESTIONS

1 Make neat, labelled diagrams of the wet sump, dry sump and petroil lubrication systems.

2 Make a list of the advantages and disadvantages of each system.

Practical: Remove engine sump, examine big end bearings, fit new sump gasket and replace sump.

21. Pumps, filters and bearings

Oil pumps

Oil is stored in the sump and pumped to various parts of the engine. Oil pumps operate all the time the engine is running and need no maintenance. They are normally driven by the camshaft or crankshaft. There are three common types:

1 Gear,
2 Rotor,
3 Eccentric vane.

GEAR TYPE

Two meshing gears – one driven by the camshaft turns the other. As they turn, oil is drawn into the space between them, carried around by the teeth and forced out when they mesh again.

Strainer

A very coarse mesh wire gauze called a strainer is fitted to the end of the oil pickup pipe in the sump to prevent large particles of carbon entering the oil system.

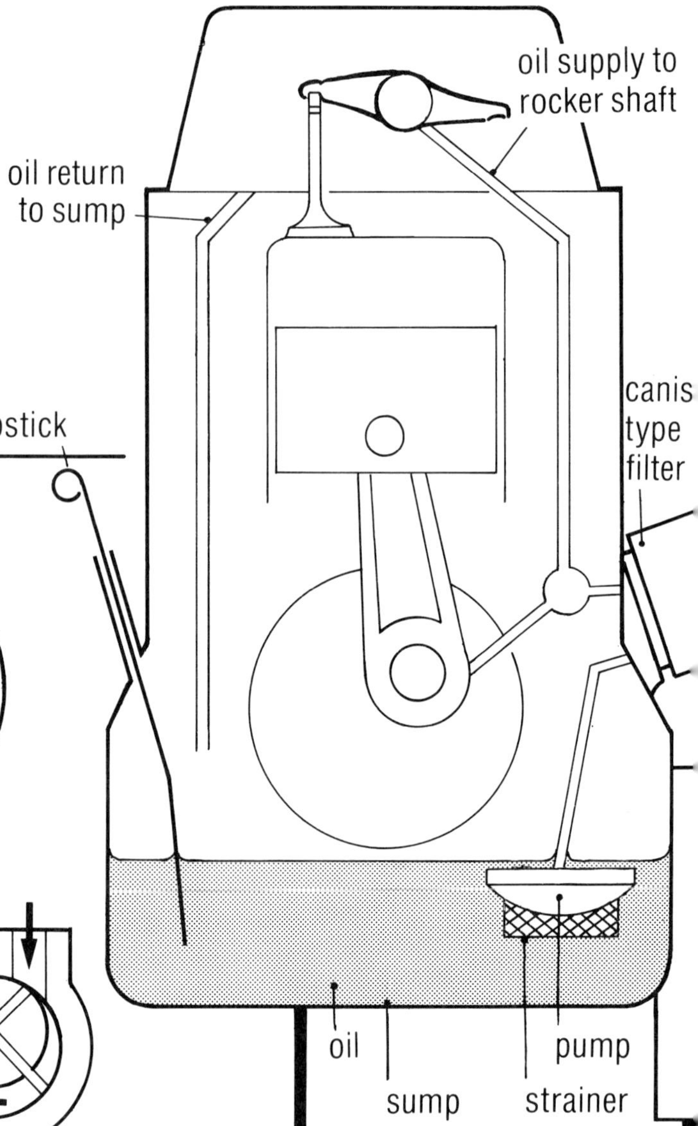

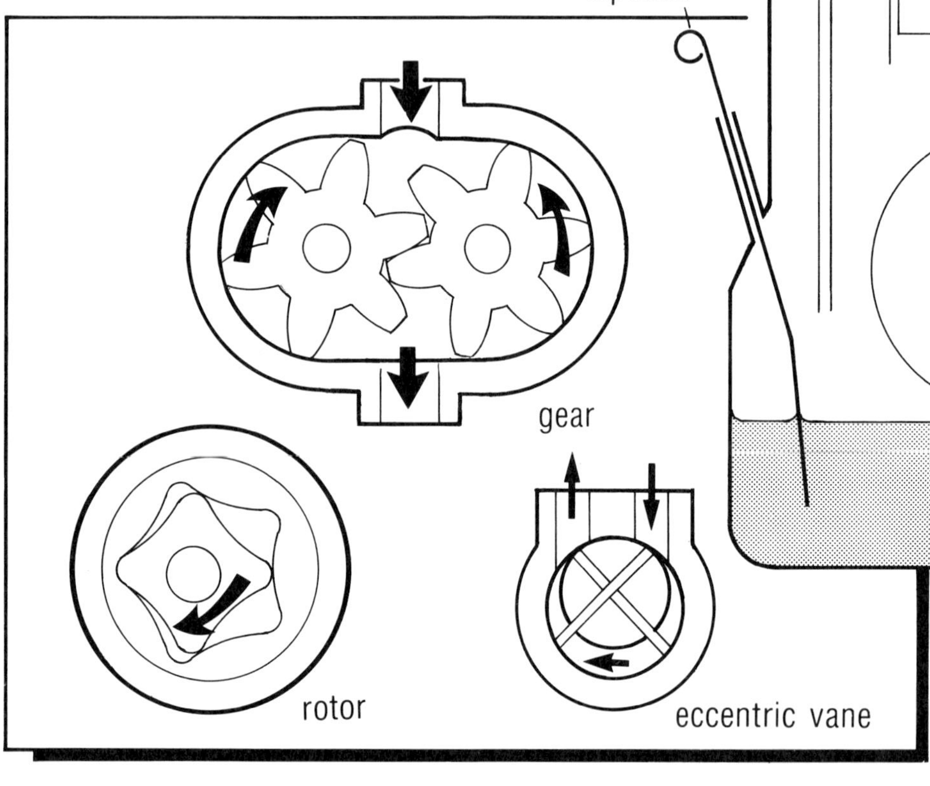

ROTOR TYPE

An inner rotor turns inside an outer rotor which has one more 'lobe' than the inner. The two rotors do not turn on the same axis, they are 'eccentric'. This means that the gaps between the rotors vary as they rotate and oil is drawn in and forced out by this motion.

ECCENTRIC VANE TYPE

A rotor with sliding vanes is mounted eccentrically in the cylinder. The movement of the vanes draws oil in through the inlet, carries it around to the outlet and forces it out.

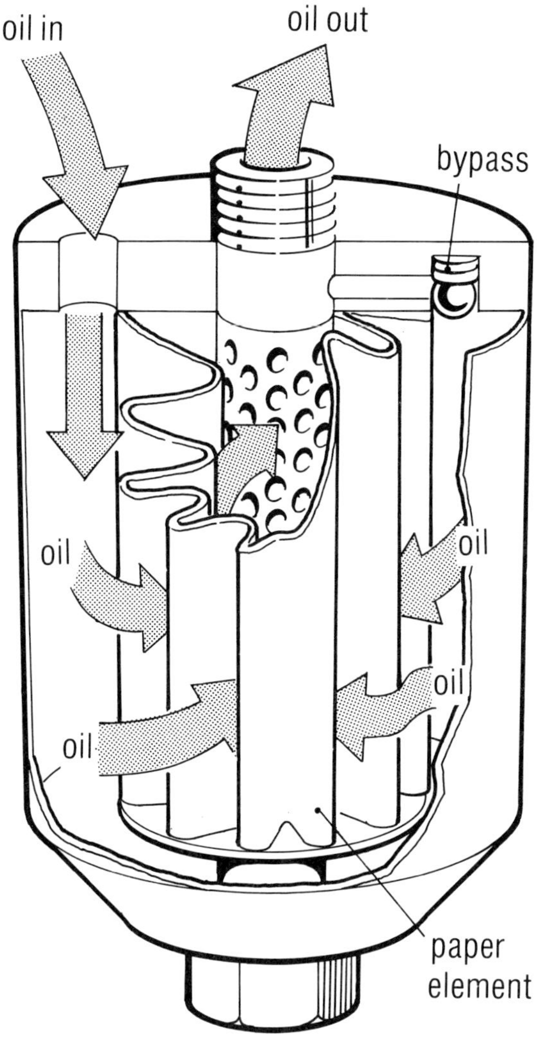

Oil filters

The main oil filter is fitted outside the crankcase. There are two types:

1 Filter bowl and paper element type,
2 Canister type.

The filter bowl type contains a paper element through which the oil will flow but impurities will not. The element can be changed quite easily when the oil is changed. The canister type screws directly onto the side of the engine and is cleaner and easier to replace, but more expensive.

If the filter becomes completely blocked it is important that the oil supply is not cut off and so a spring loaded by-pass valve is fitted. (Dirty, unfiltered oil is better than no oil at all.)

Bearings

Bearings are used to reduce friction. Bearing materials are strong (to support a load), wear-resistant and have a low coefficient of friction. Brass, bronze, nylon, white metal (an alloy of tin and copper) and aluminium alloys are all used for bearings.

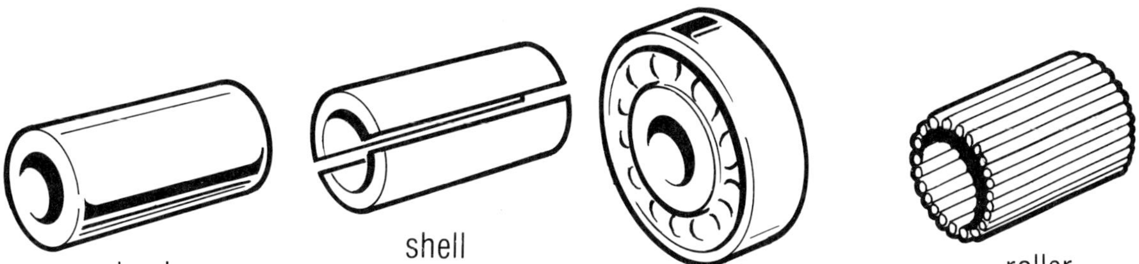

PLAIN BEARINGS
The gudgeon pin is a one-piece bearing, it is a bronze *bush*. The big end bearings are two-piece plain bearings, they are aluminium alloy *shell bearings*.

ROLLING BEARINGS
The load is taken on hardened steel rollers or balls in a steel case. Very low in friction they can take heavy loads (e.g. wheel bearings) but are expensive.

QUESTIONS

1 Make neat diagrams and notes about the three types of oil pump.
2 Write two or three sentences about each of the four main types of bearing and give an example of where each type is used.

Practical: Remove, examine and replace oil pump. Drain oil, change oil filter, run engine and check for leaks.

EXAMINATION QUESTIONS

COOLING AND LUBRICATION

1 What is the purpose of the thermostat?

2 What is the purpose of antifreeze?

3 Explain and give examples of: conduction, convection and radiation.

4 What are the advantages and disadvantages of air cooling?

5 State two reasons for using oil in an engine. (*LEA*)

6 What is meant by the term viscosity?

7 An engine repeatedly overheats within a few minutes of starting. No water leak can be detected, the cooling system is full and the fan belt tension is correct. Suggest *two* possible explanations for this problem. (*NEA*)

8 Draw three types of bearings and explain where each might be used.

9 Identify the parts indicated in the diagram of a water cooling system (Fig. 1)

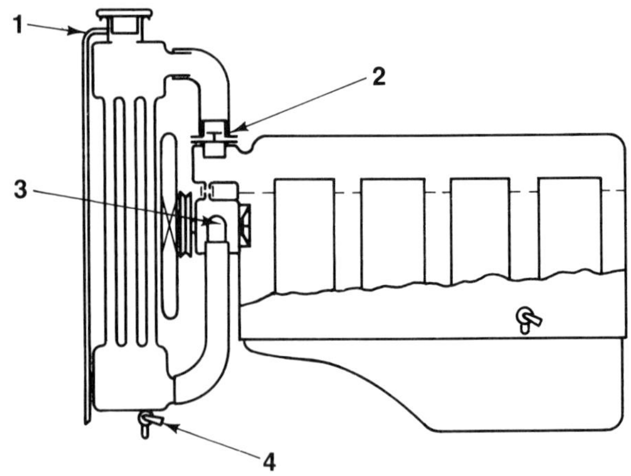

10 a) Draw a rotor type oil pump.
b) Explain in detail the petroil lubrication system.
c) What type of vehicle is likely to use a dry sump system?

Transmission

The means by which power is taken from the engine, through the clutch, gearbox and final drive, to the wheels.

22. The clutch

How the clutch works

A splined shaft, which rests in a bronze bush in the flywheel, is turned by a *friction plate.* This turns when it is pushed against the flywheel by a *pressure plate. Springs* push the pressure plate against the friction plate. When force is put on the *release bearing* several levers pull the pressure plate away from the friction plate which moves away from the flywheel and drive is disconnected. The clutch assembly is fitted inside a cover which is bolted to the flywheel and turns with it. Power is passed from the engine to the gearbox through a clutch which is bolted onto the flywheel. The clutch is used to

1 Disconnect drive from engine to gearbox for gearchanging.
2 Allow a smooth, progressive take-up of drive.

SLIPPING THE CLUTCH

As the foot is removed from the pedal the springs push the friction plate against the flywheel and drive is taken up. If the pedal is partly depressed it is possible to 'slip' the clutch so that it just touches the flywheel and turns slightly slower than the flywheel. This is useful when edging forward in traffic but causes rapid wear.

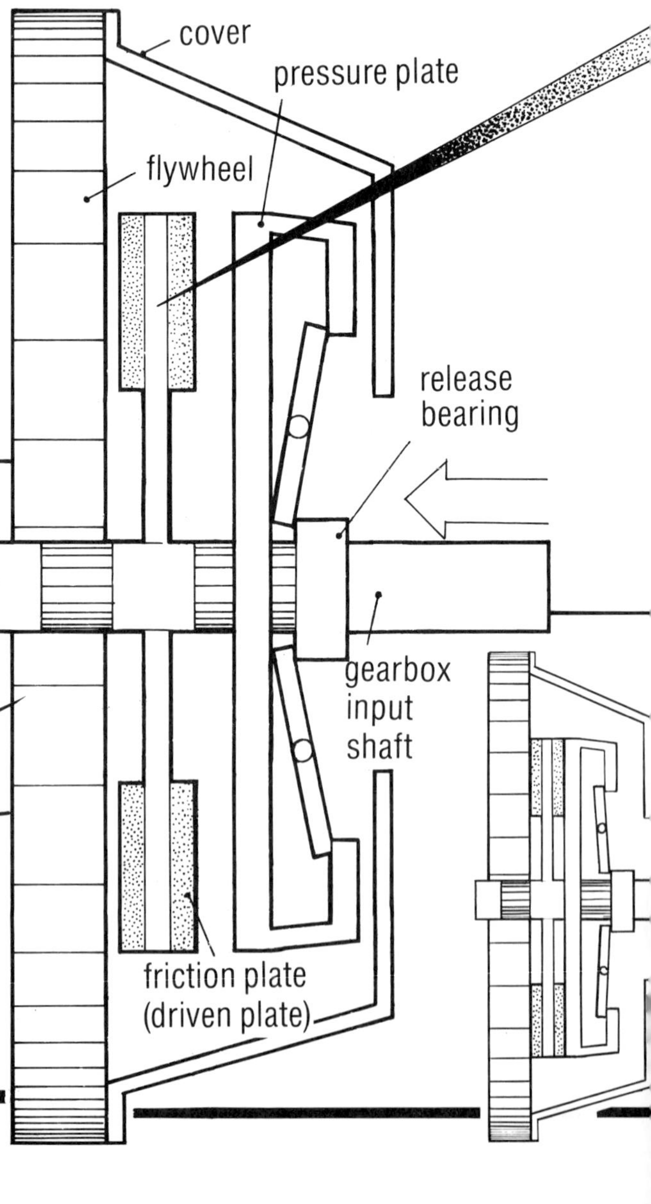

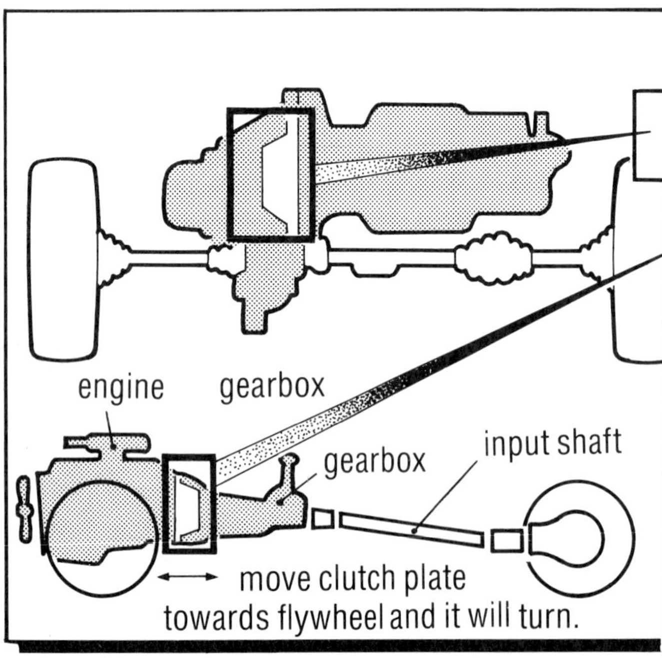

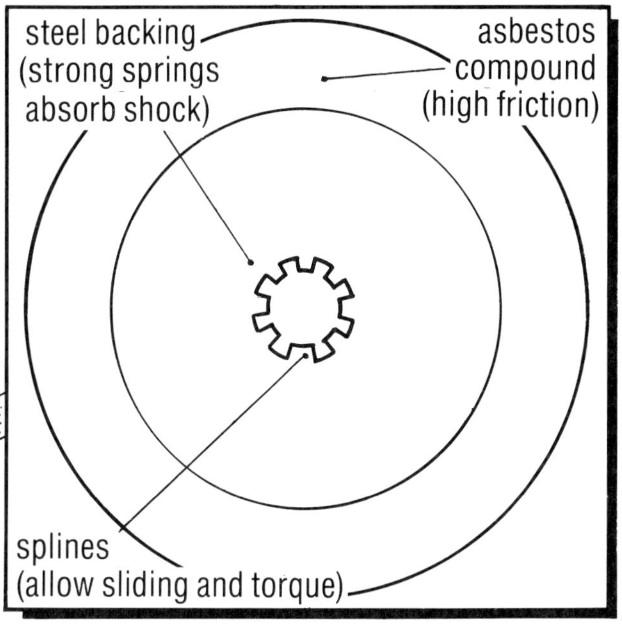

FOOT OFF PEDAL
Springs push friction plate against the flywheel, both turn.

FOOT ON PEDAL
Release bearing pushes release levers which push pressure plate back, compressing the springs. As there is no force pushing the friction plate against the flywheel the friction plate stops turning.

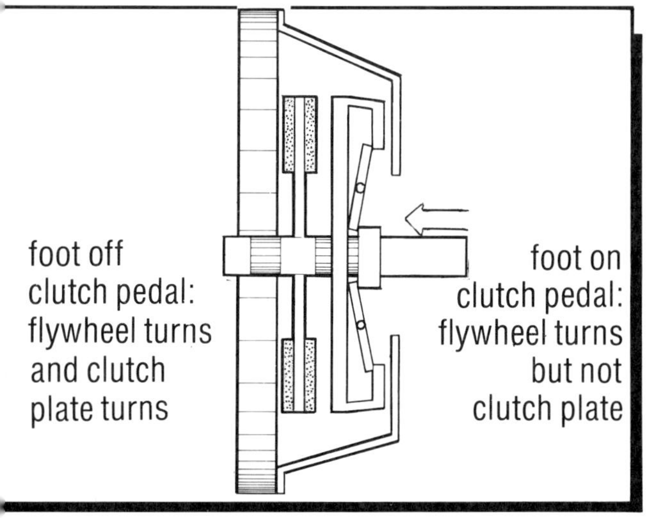

Cover assemblies

The *coil spring* is an older design with many components.

The *diaphragm spring* uses a saucer shaped spring steel plate. It is light, there are no levers, clips or separate springs and it is more efficient at high speeds.

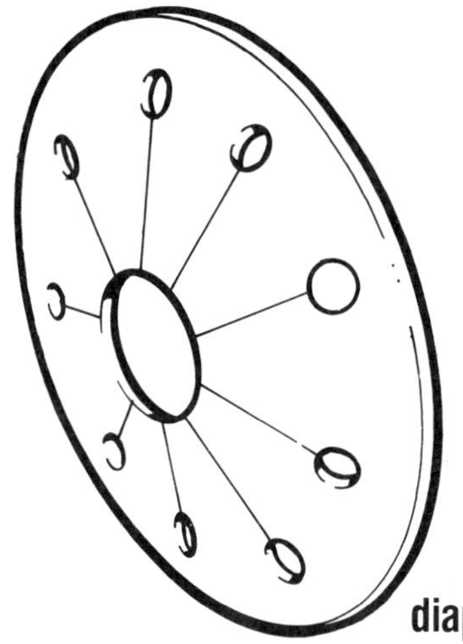

QUESTIONS

1 Explain why a clutch is necessary.
2 Make short notes and diagrams to explain how the coil spring clutch operates.

Practical: Remove clutch, check clutch plate, pressure plate, release bearing, and flywheel for wear. Replace clutch using a centring tool.

23. The gearbox

The gearbox allows the engine to run at an efficient speed for a range of different road conditions. Using toothed gears in mesh to change the speed and power of the car, the gearbox can give you a high gear for high speeds but low power (e.g. motorway driving), or it can give a low gear for low speeds but high power (e.g. starting from rest, hill climbing). It also provides a reverse gear.

Neutral: A useful position where no gears are turned on the output shaft.

First: Low speed but high power for pulling away from rest.

Second: Slightly higher speed and slightly less power. Useful for climbing hills.

Third: More speed and less power.

Fourth: Engine and propeller shaft turning at the same speed.

Reverse: Going backwards!

Types of gearbox

The input shaft turns the layshaft on which are several different sized gears. The output shaft carries different sized gears which can slide along the splines and be brought into mesh with the layshaft gears. A selector is used to move the gears. As the gears move into mesh they may be damaged and so this type of *sliding mesh gearbox* has been replaced by a design in which the gears are constantly in mesh – the *constant mesh gearbox*. Here the gears turn on the shaft but are not connected to it until a 'dog clutch' engages with it. Fitting a cone clutch to the dog clutch helps synchronise the rotation of the gears for smooth gear changes and virtually all cars now are fitted with *synchromesh gearboxes.*

Gear ratios

If two gears of the same number of teeth are put into mesh and one (the driver gear) is turned, the other gear (the driven gear) will be turned in the opposite direction at the same speed and with the same power as the driver gear.

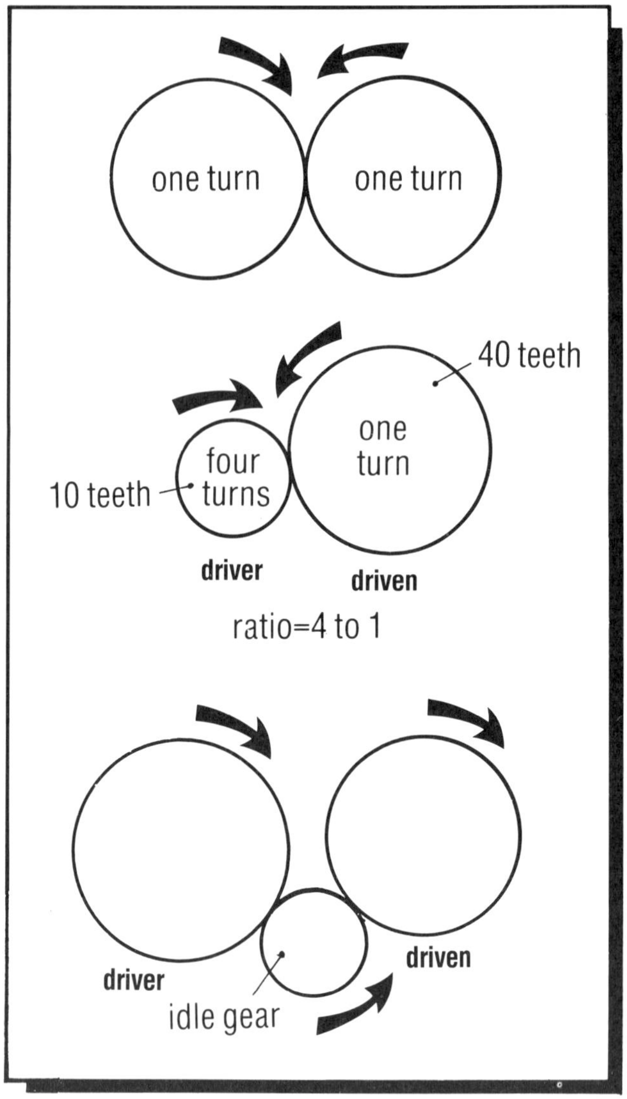

$$\text{Gear ratio} = \frac{\text{number of teeth on driven gear}}{\text{number of teeth on driver gear}}$$

If a driver gear of 60 teeth meshes with a driven gear of 30 teeth the gear ratio will be $30/60 = 1/2$ or 1 : 2. The driven gear will turn twice as fast as the driver gear but provide only half the power to the driven shaft.

A large change of speed or power requires a large gear meshing with a small gear. Large gears take up a lot of space and car gearboxes achieve large changes in gear ratios in a small gearbox by using three shafts and a gear 'train'. An *input shaft* from the engine passes power to a *layshaft* and this passes the power to the *output shaft*. The overall gear ratio is found by multiplying the ratios of each pair of gears:

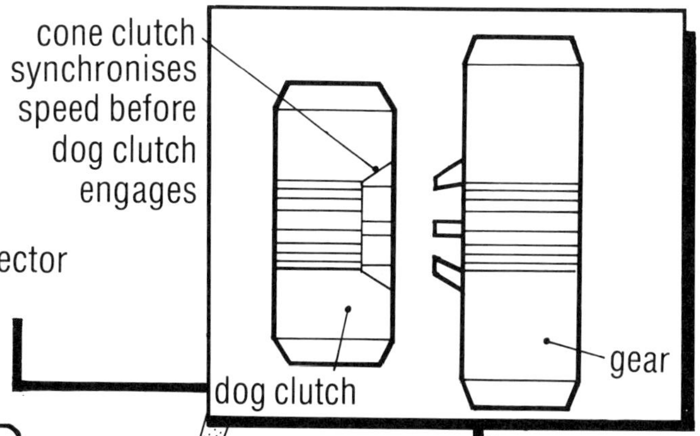

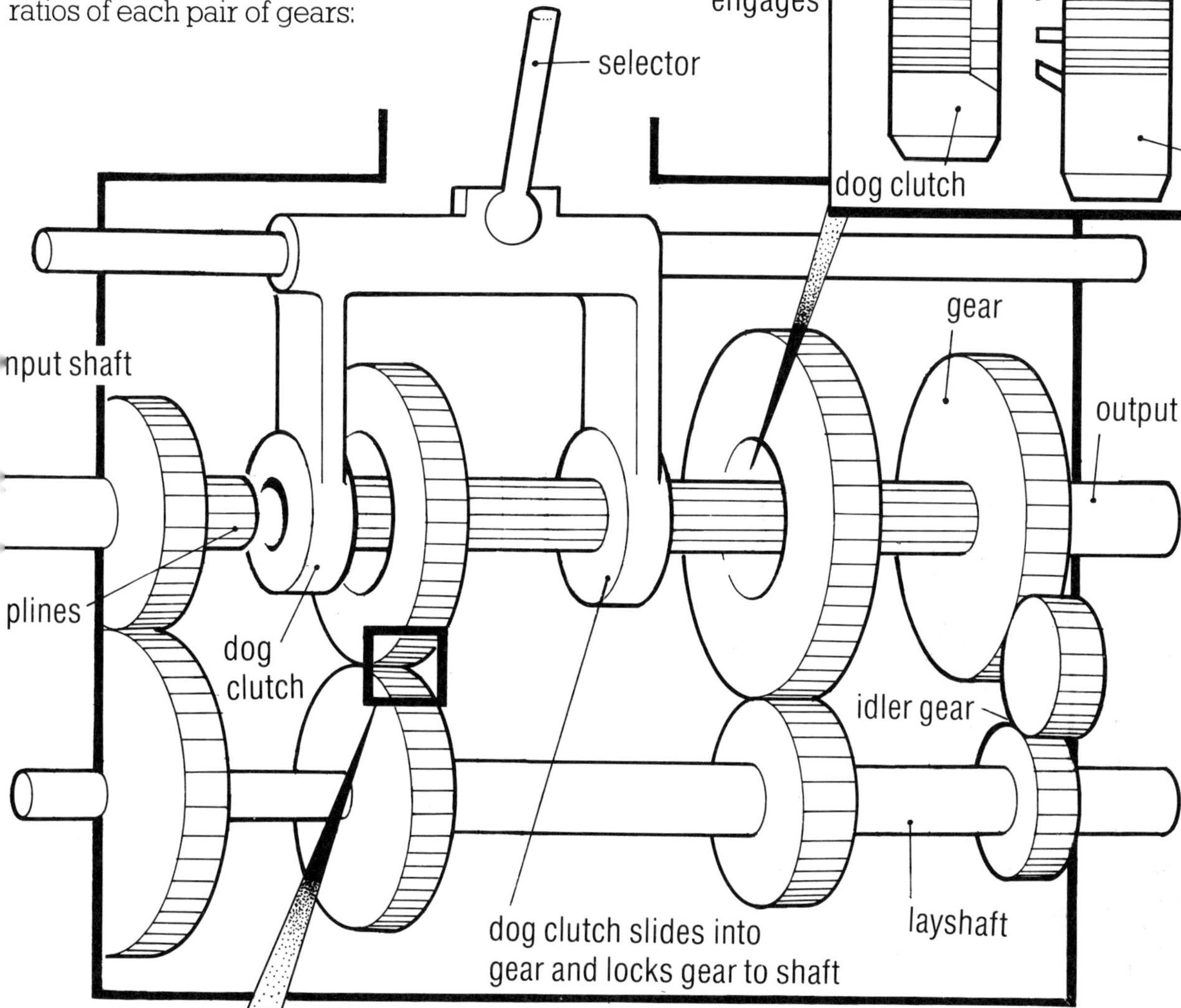

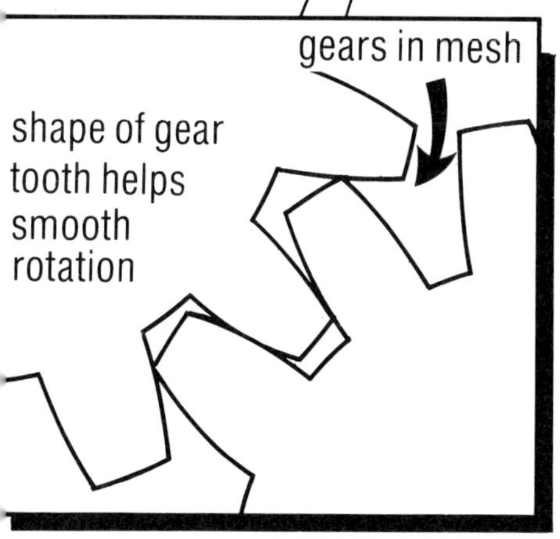

$$\text{Gear ratio} = \frac{\text{driven}}{\text{driver}} \times \frac{\text{driven}}{\text{driver}}$$

Typical gear ratios for gears 1–4 are 3.5 : 1, 2 : 1, 1.5 : 1 and 1 : 1.

REVERSE
To reverse the direction of rotation a small 'idler' gear is used. This does not affect the gear ratio.

QUESTIONS

1 Make a neat diagram to show the three gearbox shafts and write a brief explanation of how the gearbox works.

2 Calculate the gear ratios for the following pairs of gears:

Driven gear	20	20	18	54	96
Driver gear	20	10	54	18	48

Practical: Check gearbox oil level, top up if required.

24. The transmission system

The transmission system takes power from the engine and 'transmits' it to the wheels. In a front engine rear-wheel drive car each part of the system is separate. The front engine front-wheel drive has the same transmission components apart from the propeller shaft, which is replaced by drive shafts.

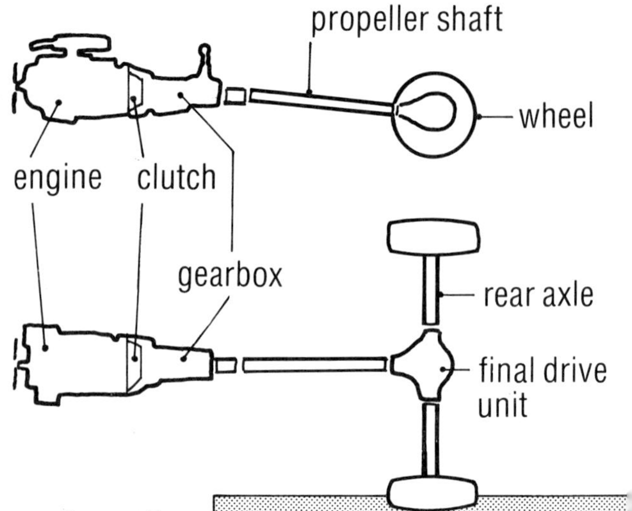

The propeller shaft

This is a hollow steel tube which makes it lighter and more rigid than a solid shaft, and less prone to 'whip' at high speeds. One end of the shaft is bolted to the final drive unit (FDU), the other end slides onto the splined gearbox output shaft.

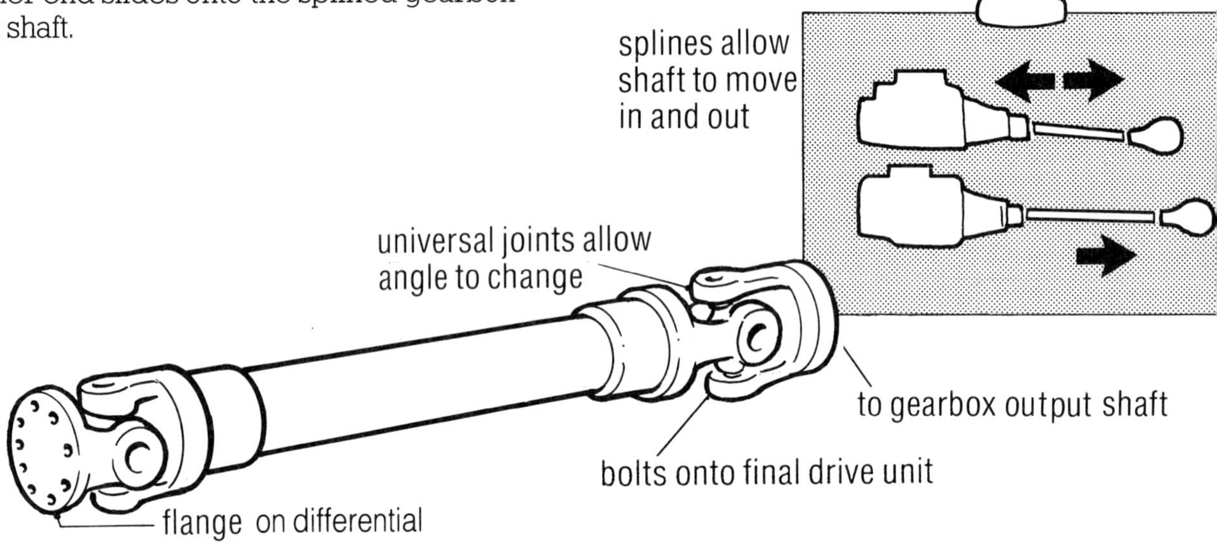

When the car goes over uneven surfaces the rear axle will move up and down. This means that the distance between the axle and the gearbox will change. The splines allow for this change in length.

The angle between the gearbox and the FDU will also change and therefore flexible joints are needed. These are called *universal joints* and are fitted to each end of the propeller shaft.

Final drive unit (FDU)

The final drive unit contains the final drive gears and the differential gears. These are very strong gears which take the turning force (*torque*) of the engine and pass it to the wheels. The FDU does three things:

1 Turns drive through 90°.
2 Gears down the final drive by about 4 : 1.
3 Provides differential gearing to allow the wheels to turn at different speeds for cornering.

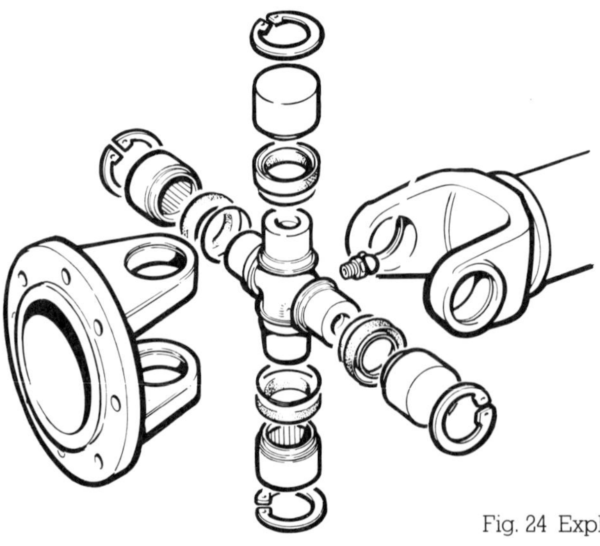

Fig. 24 Exploded diagram of universal joints

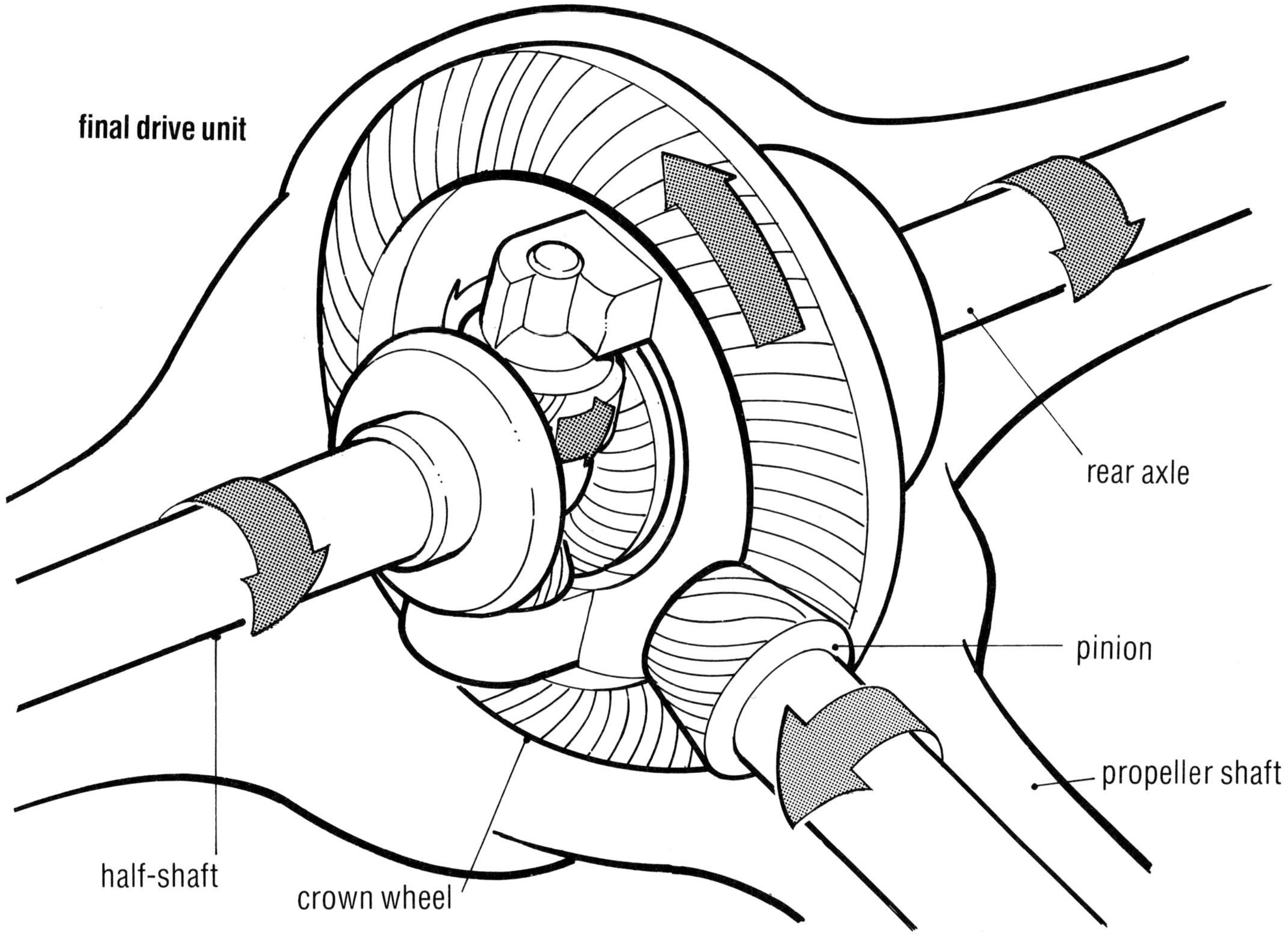

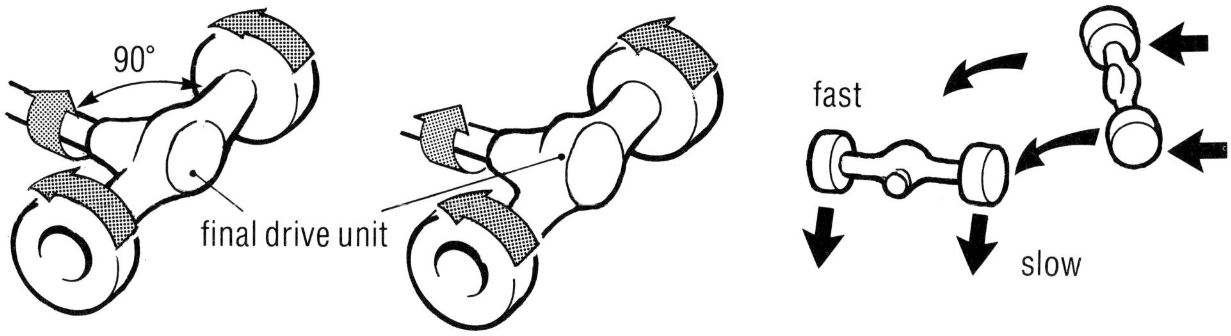

FINAL DRIVE GEARS
These are bevel gears to turn the drive through 90°, consisting of a crown wheel and pinion gear. There are more teeth on the crown wheel so that it will turn at about a quarter of the speed of the pinion.

DIFFERENTIAL GEARS
Sometimes called sun and planet gears, these are fixed to the crown wheel and turn with it. When the bevel pinions (planet gears) orbit the bevel gears (sun gears) without spinning, the wheels will turn at the same speed. When the pinions orbit and spin, the wheels will turn at different speeds.

QUESTIONS

1. Make a neat, labelled diagram of the transmission system.
2. Explain, using diagrams where necessary, what the final drive unit does in the transmission system.

Practical: Remove, examine and replace a propeller shaft.

EXAMINATION QUESTIONS

TRANSMISSION

1 State two functions of a clutch. (*LEA*)

2 Name two shafts in a motor vehicle gearbox. (*LEA*)

3 The device for disconnecting the engine from the rest of the transmission system is called:
a) universal joint;
b) clutch;
c) final drive;
d) idler sprocket? (*NEA*)

4 The single dryplate clutch as used on a modern car
a) has two driving plates, one of which is the centre-plate.
b) has one driven plate splined to the gearbox mainshaft.
c) has one driven plate splined to the gearbox primary shaft.
d) normally has a set of coil-springs in its pressure plate. (*LEA*)

5 a) Name all the numbered parts in the figure below.

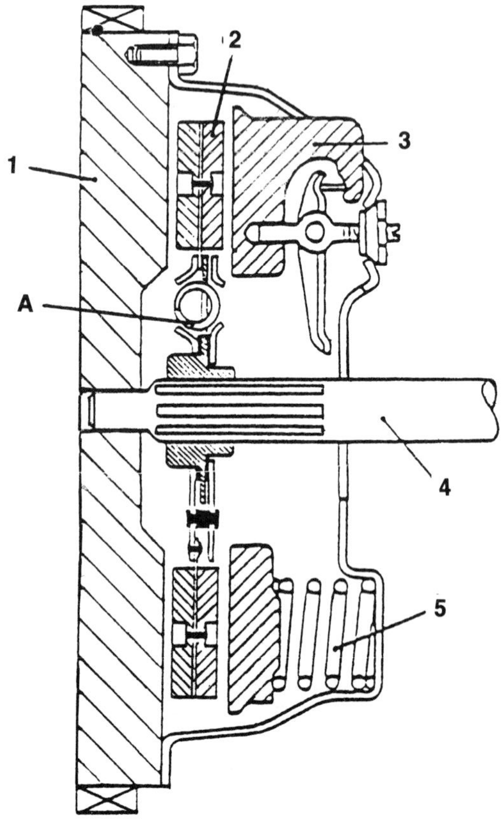

b) In which operational position is the clutch shown?
c) What is the function of component *A*?
d) Determine the effect on the pressure plate clamping force when component 2 wears.
e) Using the above diagram, explain how the drive is transmitted from engine component 1 to gearbox component 4. (*NEA*)

6 Why is a universal joint necessary on a propeller shaft?

7 Why is one end of the propeller shaft splined?

8 The diagram below shows a simplified sectioned side view of an assembly which forms a part of certain motor vehicles. Name the numbered parts. (*LEA*)

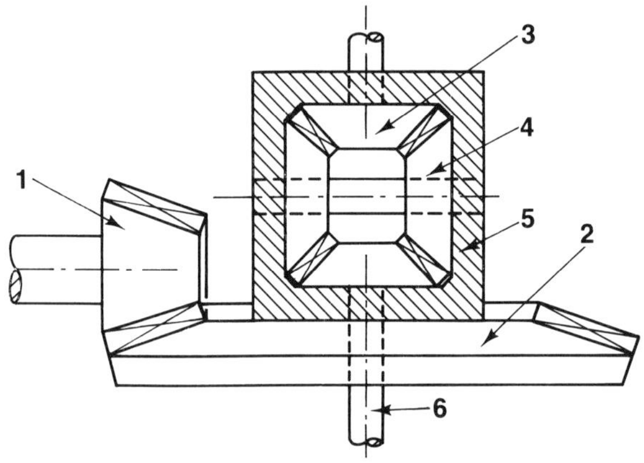

9 What is meant by the term 'friction'?

10 a) Describe briefly the purpose of a gearbox in a motor car transmission system.
b) The diagram shows three gears in mesh. If gear 'C' is turning at 600 rpm clockwise state:
(i) the speed of gear B,
(ii) the speed of gear A,
(iii) the direction of rotation of gear A.
c) If the above gears represent a typical motor vehicle gearbox layout what is the purpose of idler gear B? (*LEA*)

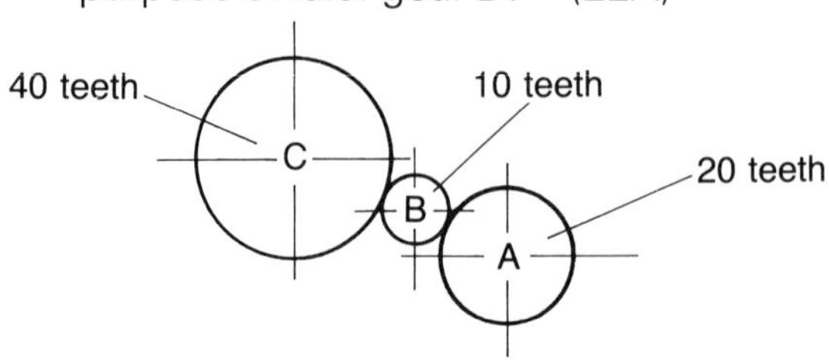

Brakes, suspension and steering

Brakes convert movement to heat. Suspension absorbs unevenness in the road surface and gives a smooth ride and improved roadholding. Steering systems allow easy, accurate control with little effort.

25. Brakes

A brake works by forcing a material with a high *coefficient of friction* against a moving disc or drum which is connected to the wheels.

Drum brakes and disc brakes

Drum brakes work by pushing a *brake shoe* against a *brake drum.* Drum brakes tend to overheat and become less efficient when used for repeated hard braking. This is called 'brake fade'.

Disc brakes work by pushing a *disc pad* against a steel *disc.* Disc brakes do not suffer from brake fade because they are exposed to the air and the flow of air keeps them cool.

BRAKE LININGS
Brake linings, made of an asbestos composition, are either riveted or bonded (glued) onto brake shoes and bonded onto disc pads.

Hydraulic braking systems

Using hydraulic fluid increases the pressure that is applied to the brakes and it is easier to route pipes than it is cables or rods. A *master cylinder* contains brake fluid which is forced along steel *brake pipes* and *flexible hoses* to the *wheel cylinders (slave cylinders).* Inside the wheel cylinder the fluid pressure forces a piston against the brake shoe, which is forced against the revolving brake drum. When the pressure is released, return springs pull the shoe away from the drum. Disc brakes do not have return springs.

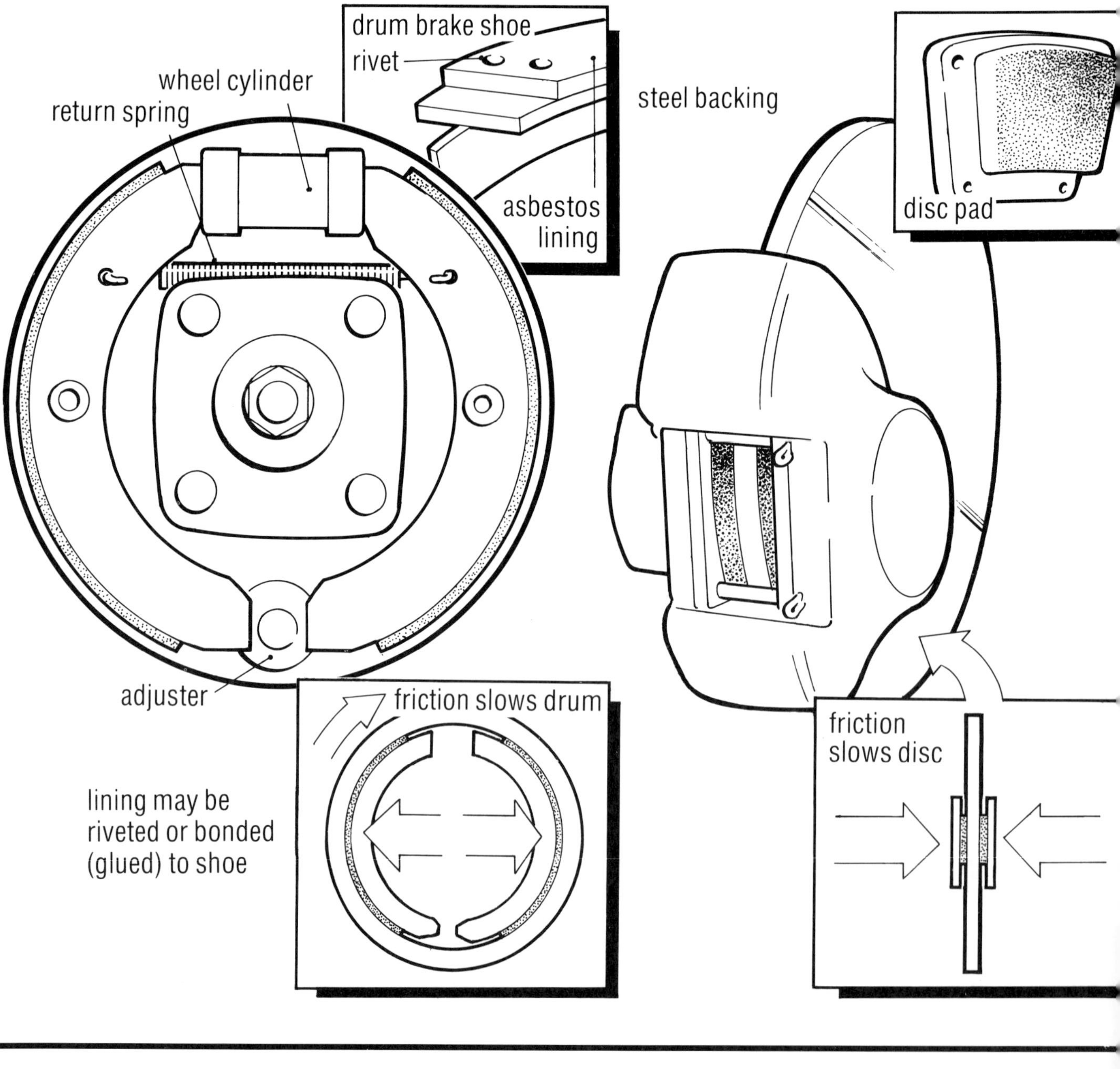

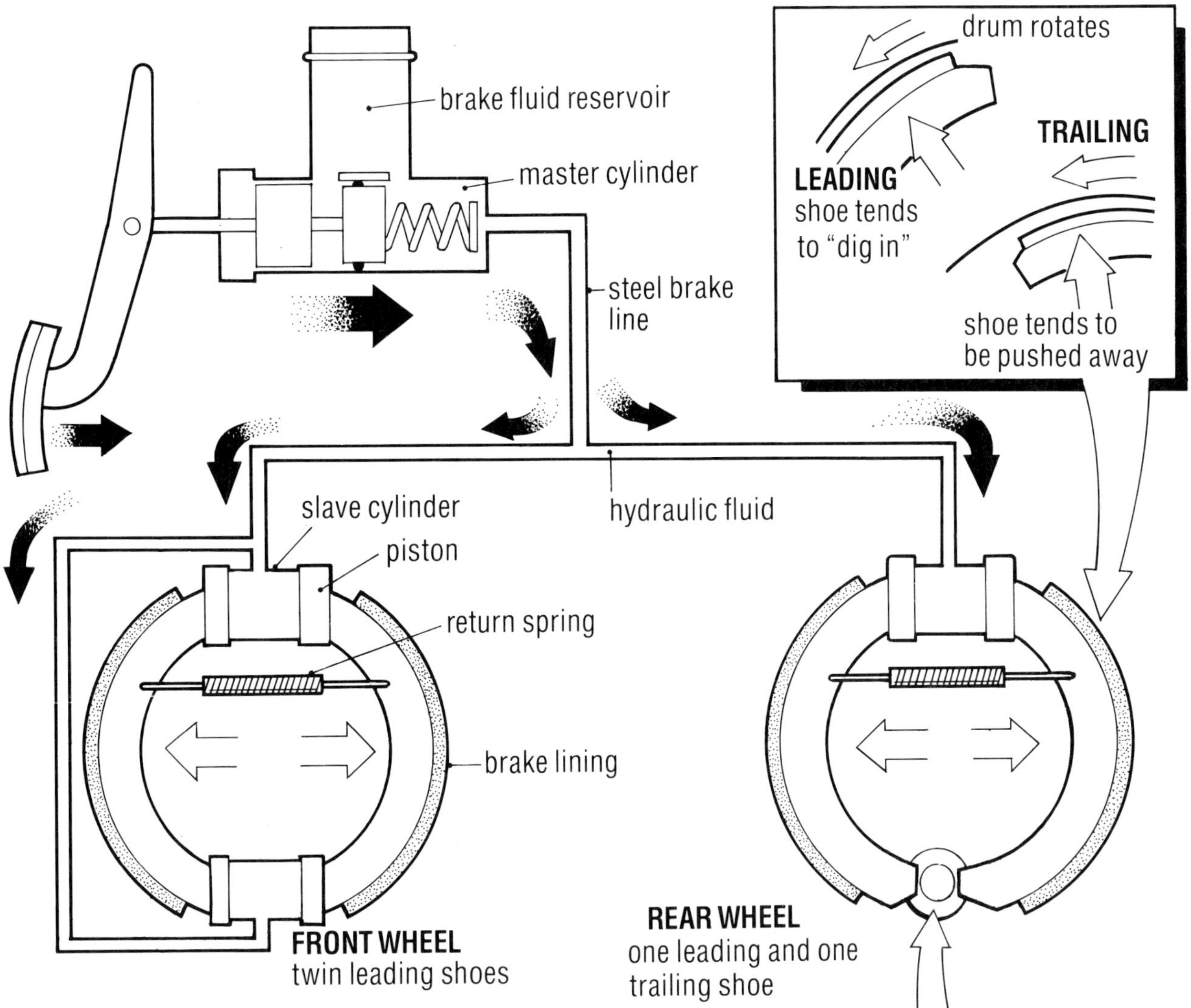

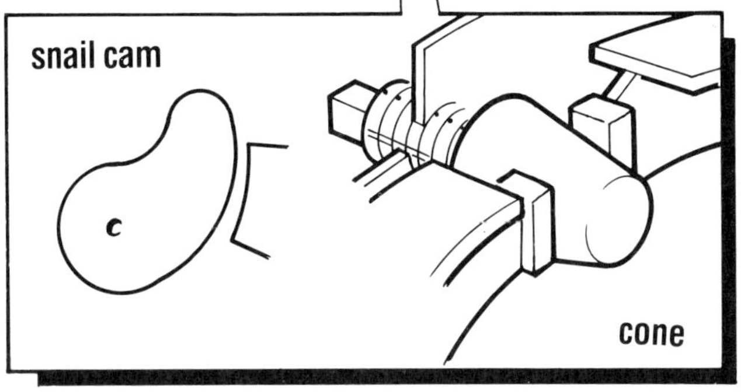

Handbrake

Cars have a mechanical handbrake to hold the car when parked and as a safety standby should the hydraulic system develop a leak. Disc brakes do not give a very efficient handbrake and many cars have disc brakes on the front and drum brakes on the rear.

Brake shoe arrangements

Leading shoe: Brake shoe tends to 'dig in'.
Trailing shoe: Brake shoe tends to be pushed away.

Twin leading shoes are fitted to the front wheels and leading/trailing shoes to the rear. When reversing, the effect is of three trailing shoes but one leading shoe, so that braking is still reasonable.

Brake adjusters

Snail cam: Turning cam moves shoe closer to drum.
Cone: Turning adjuster moves cone inwards, which moves shoe closer to drum.

QUESTIONS

1 Make a neat, labelled diagram of a hydraulic braking system.
2 Explain with diagrams and short notes the differences between disc brakes and drum brakes. Mention the advantages and disadvantages of each type of brake.

Practical: Check brake fluid level in master cylinder and top up if necessary.

26. Wheels and tyres

Early wheels were wooden with steel rims. They were noisy, uncomfortable, had poor grip and put a strain on the vehicle. Solid rubber tyres were an improvement but the air filled (pneumatic) tyres developed by John Dunlop gave a much smoother ride and better grip.

Tyres need vertical flexibility for a comfortable ride but lateral (side to side) rigidity for good steering and roadholding. Smooth tyres (slicks) give a good grip on a smooth racetrack, but not on ordinary roads in the wet. The tread on a tyre improves roadholding on a wet road by dispersing the water.

Types of tyre

There are two main types of tyre used in cars today: cross-ply and radial-ply.

Do not mix tyres. Cross-ply and radial-ply tyres have different handling characteristics and mixing tyres on the same axle may lead to the car going out of control when cornering. It is best to fit radials all round, but it is permissible to fit radials to the rear and cross-ply to the front axle.

TUBED OR TUBELESS

Both types of tyres may be either fitted with an inner tube (as on a bicycle) or without a tube. Tubeless tyres, used on most cars, are easier to fit, they hold air longer and do not go down so quickly if punctured. Tubes are needed with wire spoked wheels.

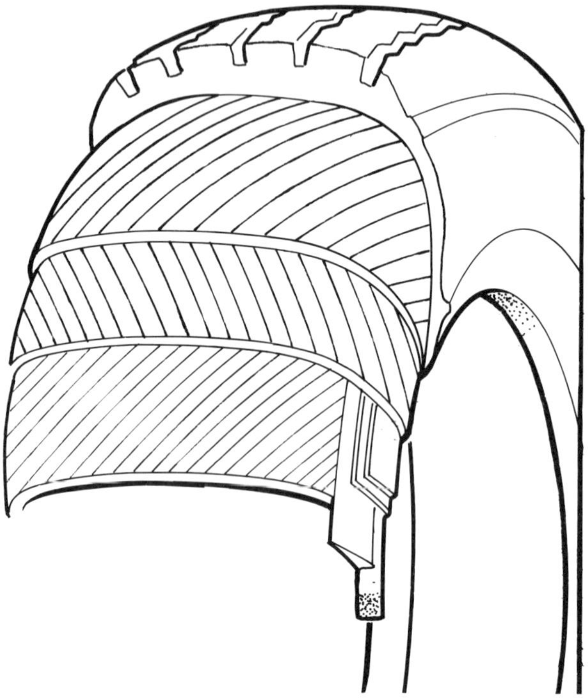

Cross Ply

Ply means layer. Plies are put on top of one another at an angle.

Good comfort but not very good cornering.

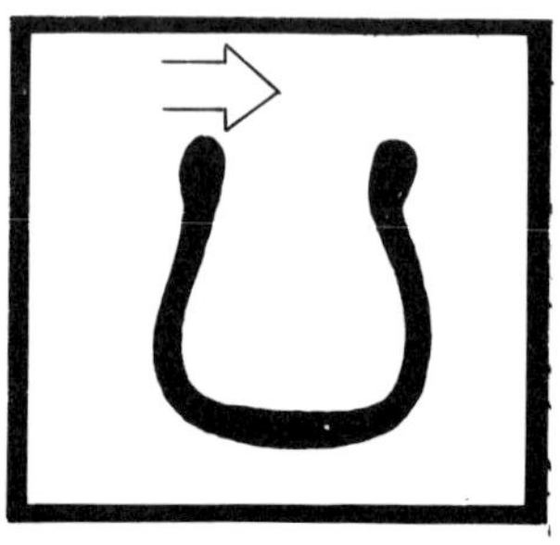

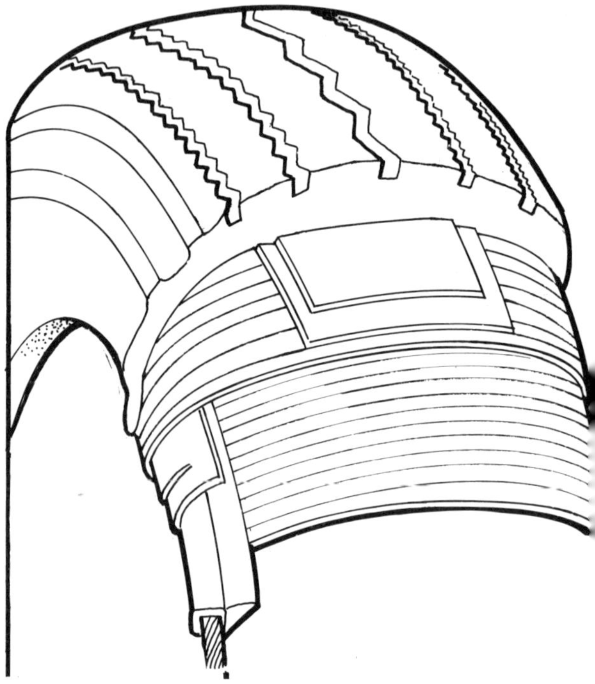

Radial ply

Plies are at right angles to wheel with a strengthening layer over them.

More flexible side wall than cross-ply so it corners better, lasts longer but is more expensive, slig noisier and steering is heavier at low speeds.

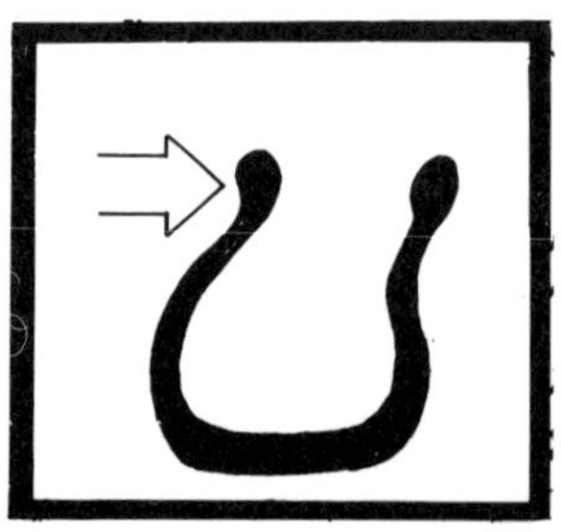

Wheel rims

The wheel rim is designed to hold the tyre. Air pressure pushes the *tyre bead* against the rim. The *well* makes removal of the tyre easier. Some specialist rims may be split rims bolted together.

RIM DIAMETER

Numbers stamped on the tyre (e.g. 165 × 13 or 145 × 12) refer to the width and diameter of the tyre. The first number is the tyre width in millimetres. (Fatter tyres give better roadholding but cost more; larger cars need larger tyres.) The second number is the rim diameter in inches.

Fig 26(a) Effects of over inflation

Fig 26(b) Effects of under inflation

Fig 26(c) Incorrect 'tracking'

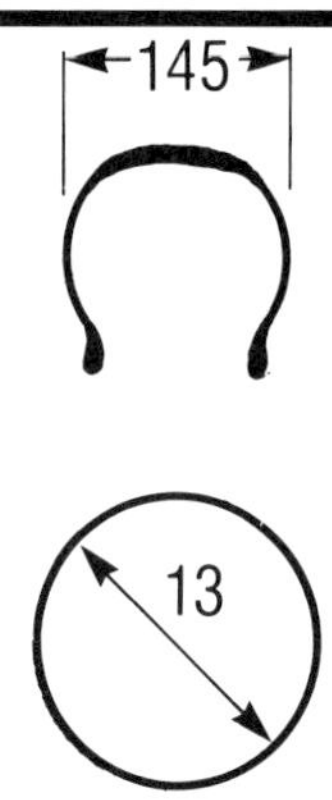

Three types of wheel

Pressed steel used on most cars, cheap and reasonably strong. Usually decorated with a chrome plated wheel trim.

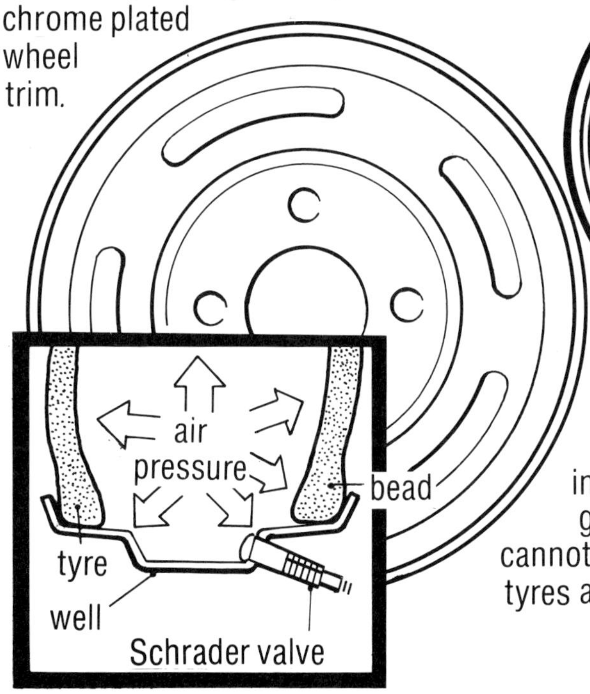

Alloy lighter, better roadholding, attractive, expensive. Magnesium alloy is light but strong.

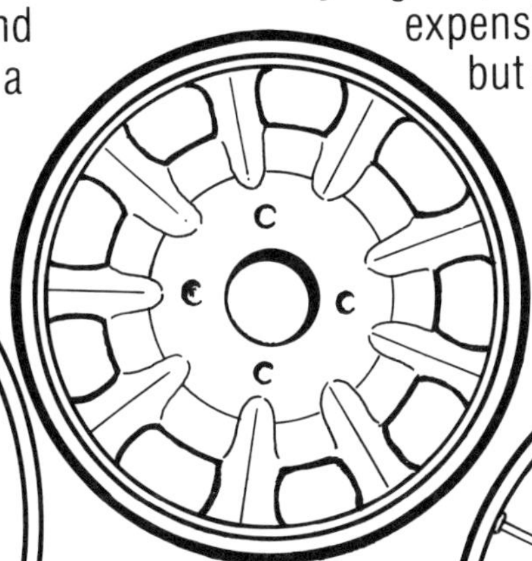

Wire spoked used on sports cars, interesting appearance, good airflow to brakes, cannot be used with tubeless tyres and are very difficult to clean.

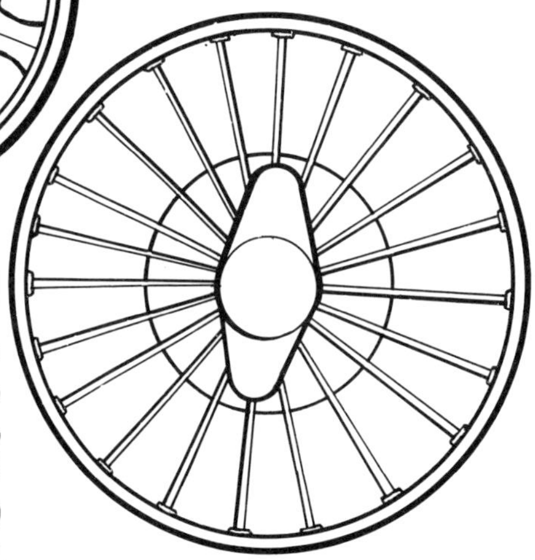

QUESTIONS

1 Explain, using diagrams and short notes, the difference between cross-ply and radial-ply tyres.

2 List the advantages and disadvantages of each of the three types of wheels.

Practical: Jack up car, put onto axle stands, remove wheel, examine tyre for damage, check tread depth. Replace wheel, remove axle stands and lower car to the ground.

27. Suspension

Suspension absorbs bumps and dips in the road surface and results in:

1 A smoother ride (because the car moves less).
2 Safer handling (because the wheels stay on the road).
3 Less mechanical wear (because the car is shaken about less).

Rear suspension

LEAF SPRING

A simple, common rear suspension system uses a semi-elliptic leaf spring. Thin strips of spring steel clamped together form the spring. There is a swinging shackle at one end to allow for the lengthening and shortening of the spring as it compresses and rebounds. The layout of propellor shaft, solid rear axle and leaf springs is called a Hotchkiss drive.

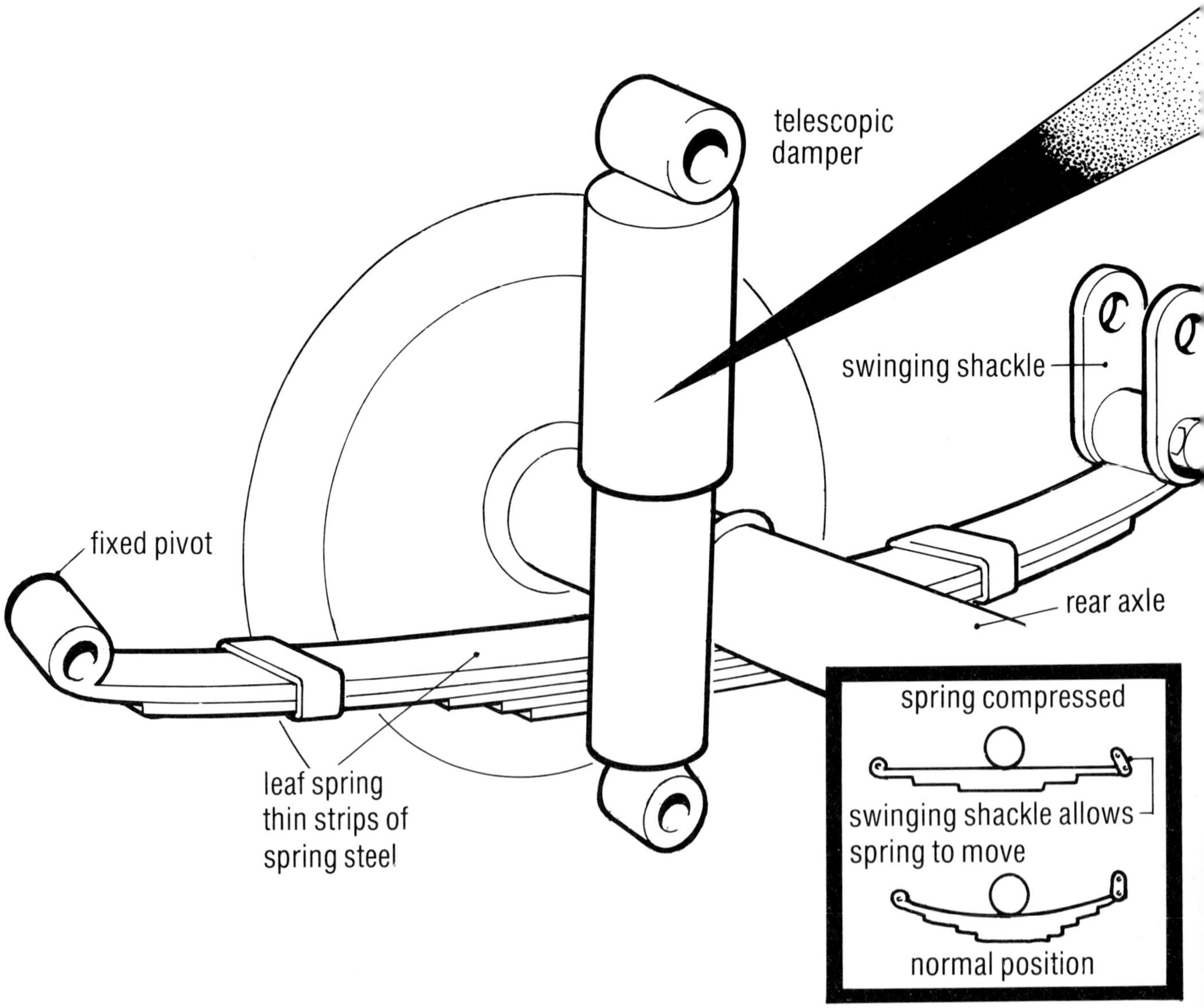

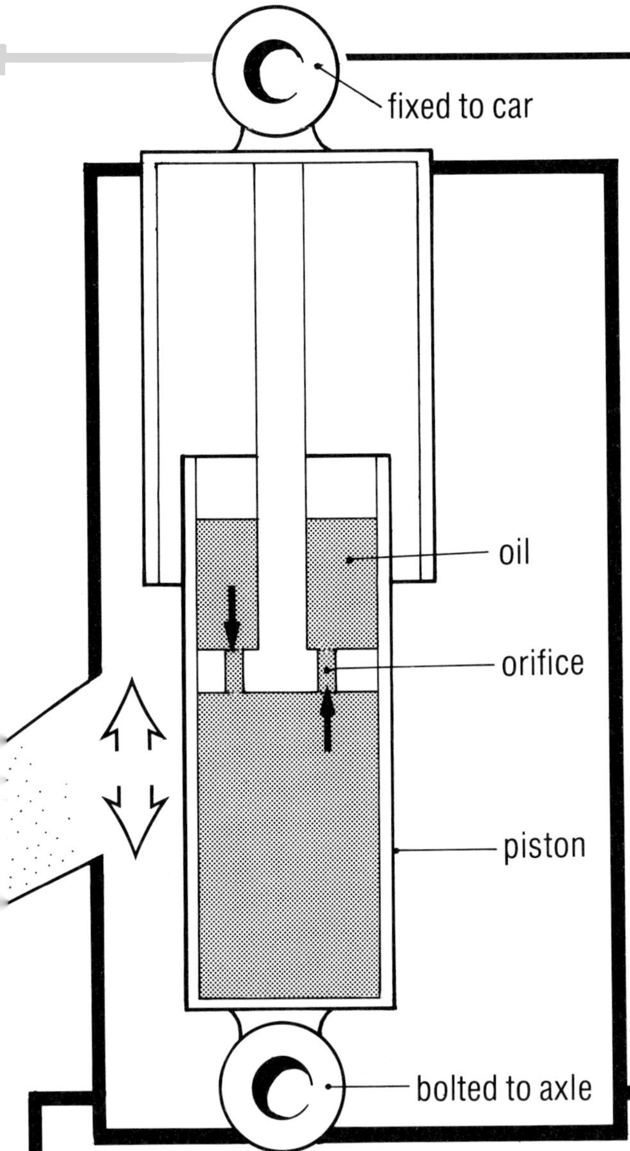

INDEPENDENT REAR SUSPENSION
Independent suspension on each of the wheels is better because movement in one does not affect the other as it does in the solid axle, leaf spring system. It is, however, more complicated (and therefore more expensive!) because the axle must be split and two drive shafts are necessary.

DAMPERS
A problem with springs is that once they have been compressed they tend to spring back!

Dampers slow down the action of the springs and make things smoother. As the car moves, a piston with a small hole in it moves through oil and the action of the suspension is slowed. The movement of the car is 'damped' and spring bounce is reduced.

Front suspension

There are two main types of independent front suspension:

1 McPherson strut which is simple and quite effective.
2 Double wishbone which is more compact.

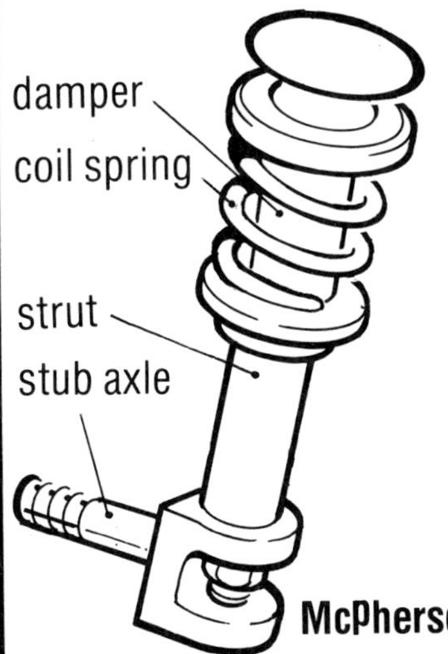

McPherson strut

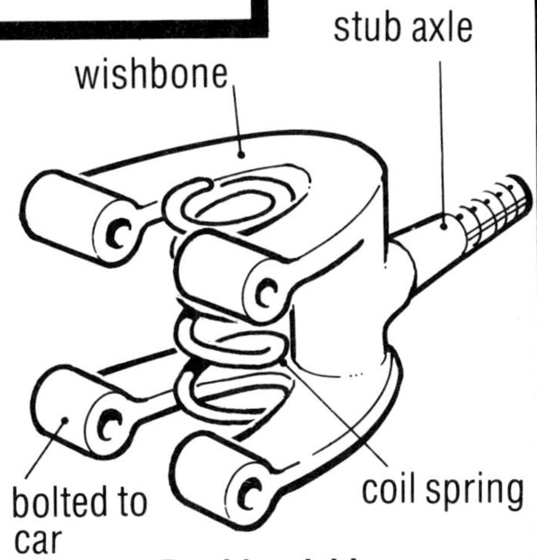

Double wishbone

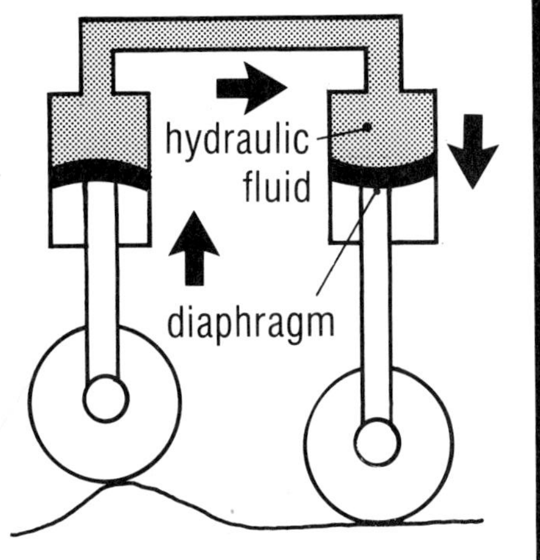

Hydrolastic suspension

Hydraulic suspension

Hydraulic fluid moves through pipes between cylinders on each wheel. This keeps the car level and reduces the tendency of the car to 'pitch' as it goes over a bump.

QUESTIONS

1 Describe, with diagrams and short notes, the two main types of front suspension and rear suspension

2 Explain why a suspension system is necessary on a car.

28. Steering

The steering system has to enable the driver to control the direction of the car quickly and easily at all speeds. Each wheel turns on its own short *stub axles*. If the wheels were fixed to a solid axle (as in early horse drawn vehicles) there would be a very large turning circle and the bodywork would have to be very high or have cutouts.

When a car turns a corner the inner wheel travels a shorter distance than the outer wheel and so has to be at a different angle. If each wheel turns on its own stub axle and is controlled by a track rod this is possible. The angle between the king pin and the track rod joint is such that, if it were continued, it would meet the back axle in the centre. This layout is based on the *Ackermann principle* and the angle is called the *Ackermann angle*.

Steering systems

To reduce the effort needed to turn the car the steering system is geared. A rack and pinion gear will turn the rotation of the steering wheel to straight line (linear) movement for the track rod. Another type of gear is the screw and nut and, similar to this, the cam and peg. The important thing is to provide easy, positive steering that allows the driver to 'feel' what is happening to the wheels without passing back too many road shocks.

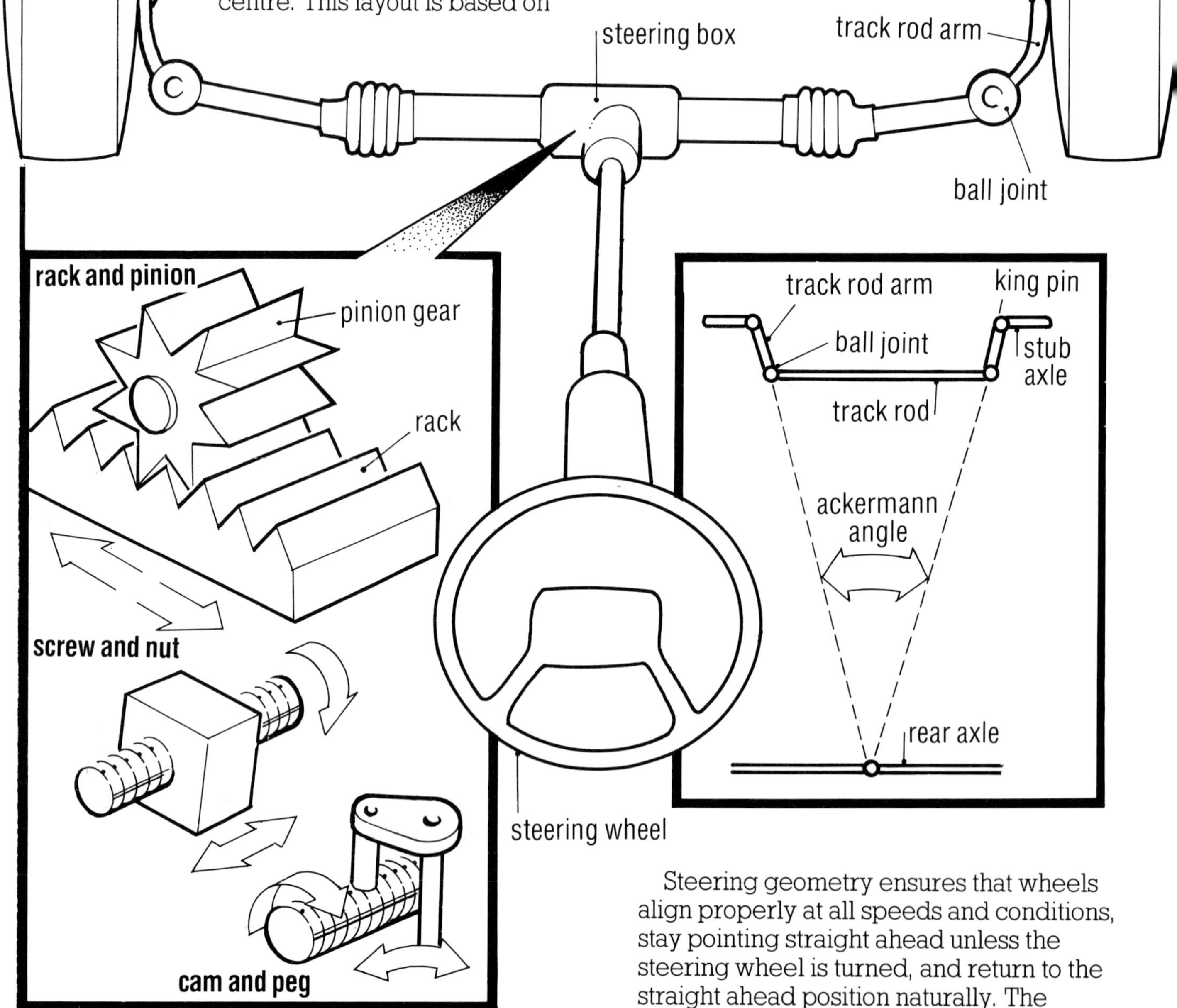

Steering geometry ensures that wheels align properly at all speeds and conditions, stay pointing straight ahead unless the steering wheel is turned, and return to the straight ahead position naturally. The

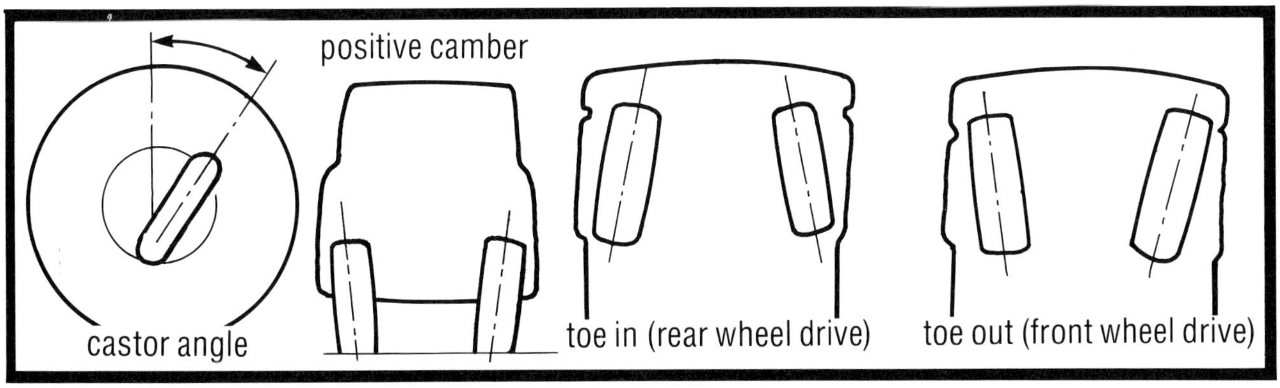

geometry involved is quite complex but these are the main points.

CASTOR ANGLE
This ensures that the wheels point straight ahead unless the steering wheel is turned. It is not adjustable. [The wheels on supermarket trolleys are castors.]

CAMBER ANGLE
To improve roadholding and give easy steering even when loaded the wheels are set at a slight angle to the vertical. Most cars have positive camber but some sports cars use negative camber. Roads have a slight camber.

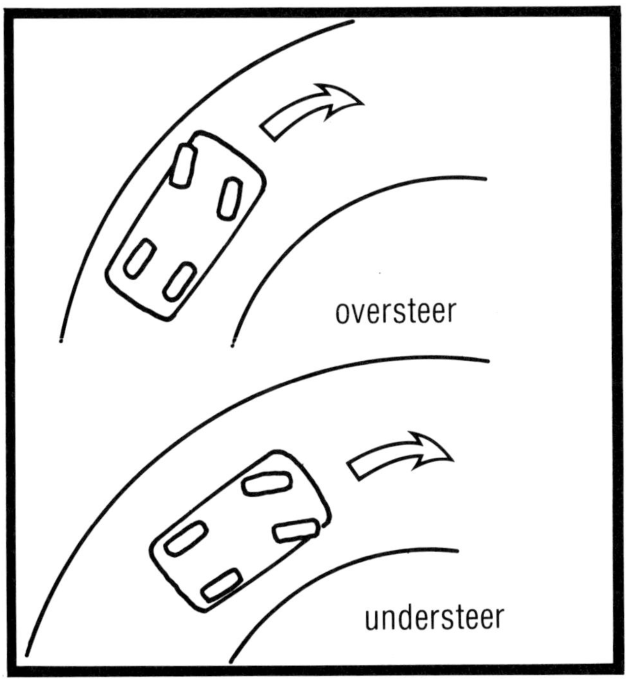

'TRACKING'
Wheels should be parallel when the car is moving but because there is 'play' in the system, wheel alignment is set slightly out of parallel when at rest. This angle is adjustable and keeps the car on the right track.

Understeer and oversteer

If a car is cornered very fast, one end of the car will tend to slip before the other. Either the front wheels will tend to lose grip and slip first (understeer) or the rear (oversteer). Most cars are designed to give slight understeer.

Collapsible steering column

A solid shaft connected to the wheel is dangerous in an accident and so many cars are fitted with collapsible steering columns.

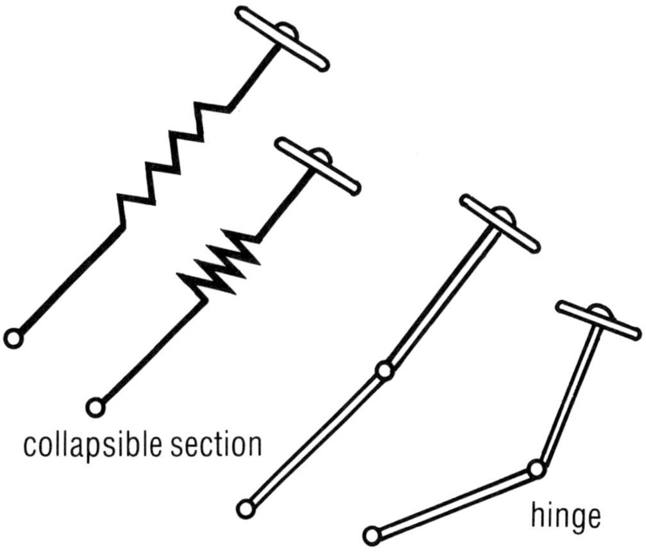

QUESTIONS

1 Make a neat; labelled drawing of a steering system, showing king pin, stub axle, track arm and Ackermann angle.

2 Write two or three sentences about each of the main angles of steering geometry.

Practical: Examine steering joints and connections for wear. Set 'tracking'.

EXAMINATION QUESTIONS

BRAKES, SUSPENSION AND STEERING

1 What is meant by the term 'brake fade'?

2 State the two types of brake used on a modern motor car. (*LEA*)

3 Which one of the following spring types lends itself to the design of a suspension system having a self-levelling ability?
a) Leaf,
b) Coil,
c) Gas,
d) Rubber. (*LEA*)

4 "When the vehicle strikes a bump in the road the rear of the vehicle bounces up and down for a considerable time. I lubricated the leaf springs and this seems to have increased the bounce."

State:
a) a possible cause;
b) briefly how you would rectify the fault. (*NEA*)

5 The 'feathering' type of wear occurring on both front tyres could be caused by
a) excessive front wheel toe-in.
b) too much positive camber on the nearside wheel.
c) over inflation on both front wheels.
d) zero castor on both front wheels. (*LEA*)

6 Which of the following tyre arrangements is allowable on cars?
a) Crossply on front, radial on rear;
b) Radial on front, crossply on rear;
c) Crossply on nearside, radial on offside;
d) Radial on nearside, crossply on offside. (*LEA*)

7 The diagram shows a side view of one side of a particular type of independent front suspension. Name the numbered parts. (*LEA*)

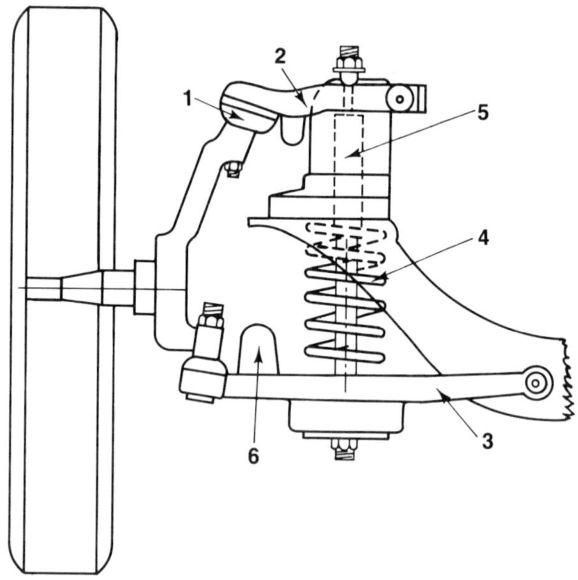

8 State *four* factors related to tyre construction and condition which affect tyre adhesion. (*NEA*)

9 Explain, with relevant diagrams and short notes, the following terms:
a) camber angle
b) castor angle
c) Ackermann angle
d) understeer
e) toe in.

10 a) What is meant by 'leading' and 'trailing' shoes in a drum brake?
b) What advantage has the disc brake when compared with a drum brake?
c) With the aid of a simple diagram describe the method of operation of a disc brake. (*LEA*)

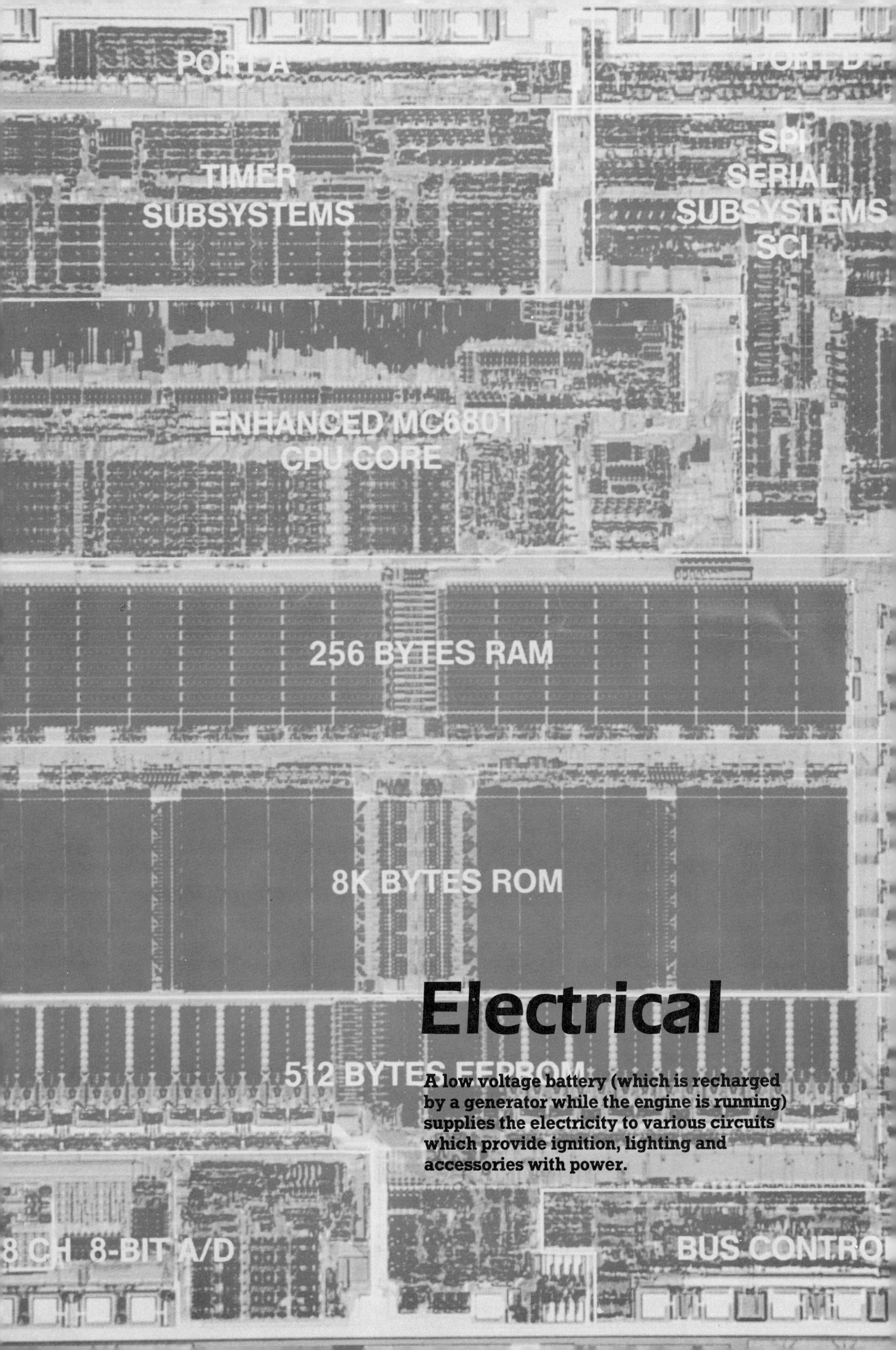

Electrical

A low voltage battery (which is recharged by a generator while the engine is running) supplies the electricity to various circuits which provide ignition, lighting and accessories with power.

29. Electrical circuits

Electrical components are joined together to make circuits. The main circuits in a car are: lights, ignition, starting, charging and accessories. To draw a circuit diagram by drawing the actual components would take a long time and be quite complex and so symbols are used. The illustration (right) shows some of the more common symbols used in circuit diagrams for car electrical circuits.

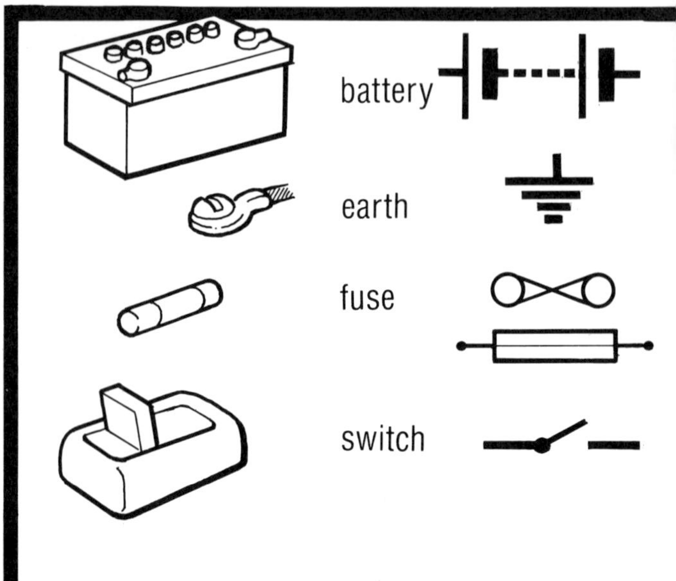

Two types of circuit

1 *Series:* If one bulb breaks the others will also go out.
2 *Parallel:* If one bulb breaks the others are not affected.

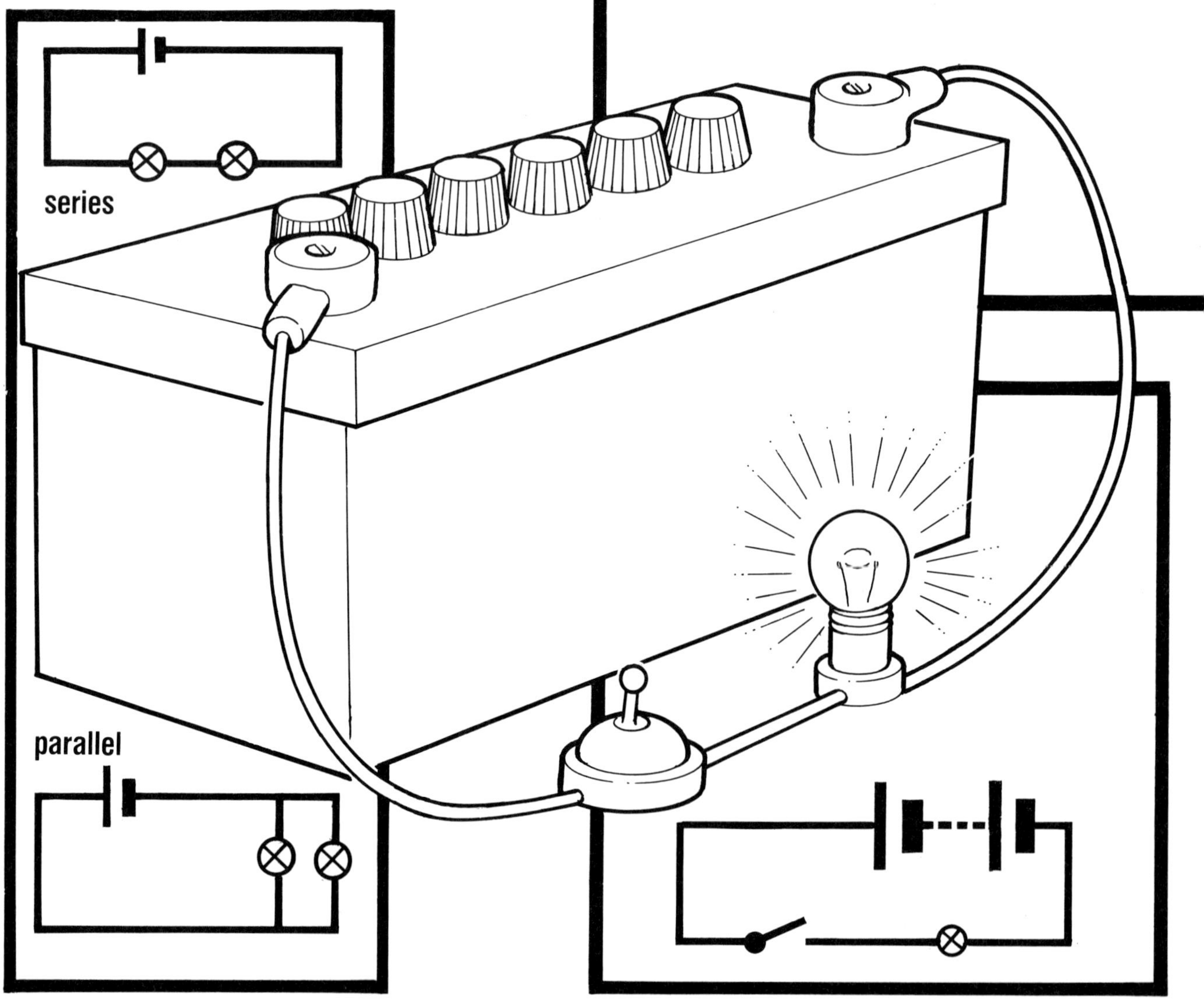

Fuse

A safety device, like a weak link in a chain. If too much current flows or if a wire becomes loose and causes a short circuit the thin wire inside the fuse overheats and melts, breaking the circuit. The fuse protects other components from damage.

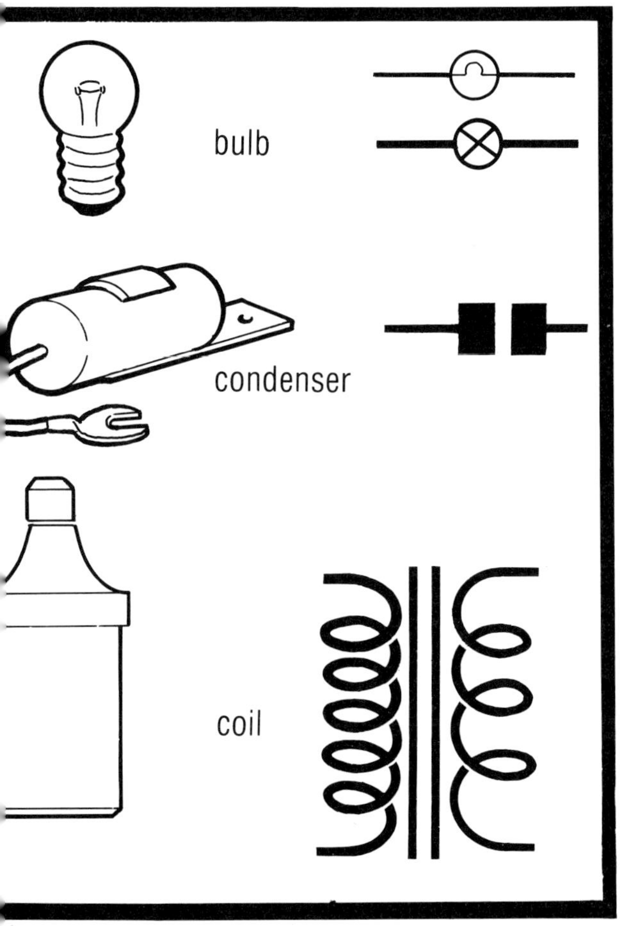

Colour code

To make it easier to trace wires, different circuits use different coloured wires.

Brown: battery and generator
White: ignition
Blue: headlamps
Red: side and tail lights
Black: earth
Green: fused circuits from the ignition switch
Purple: fused circuits not from the ignition switch.

connectors

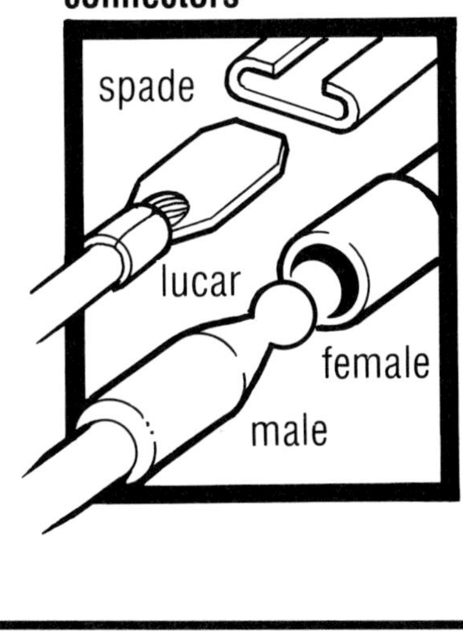

Conductors and insulators

Materials which allow electricity to pass through them are called *conductors*. Most metals are conductors. Copper is a very good conductor. Materials which prevent electricity flowing are called *insulators*. Plastics are insulators, as are rubber, nylon, glass and ceramic.

Earth return

One terminal of the battery is connected to the body of the car and electricity returns from the components to the battery by this route. This saves wire and also reduces the number of connections in the circuits. Modern cars have a negative earth as this causes less corrosion than a positive earth system.

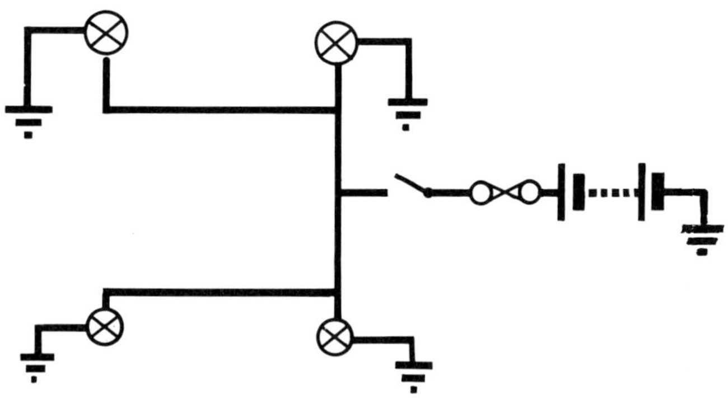

A simple lighting circuit

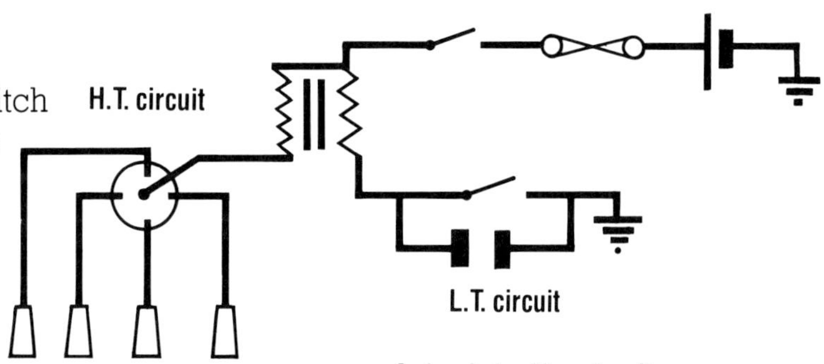

A simple ignition circuit

QUESTIONS

1 Make a neat, labelled circuit diagram of the ignition circuit.
2 Write two or three sentences to explain each of the following terms: conductor, insulator, fuse, earth return, series circuit, parallel circuit.

Practical: Make a simple lighting circuit.

30. Electrical components

Starter motor

Electricity passing through the coils of wire on the armature make them magnetic. They cause the armature to rotate as they are attracted and repelled by the magnetic field coils. As the shaft turns at very high speed the pinion gear moves along a thread on the shaft and engages with the starter ring gear on the flywheel. This turns the engine over for starting.

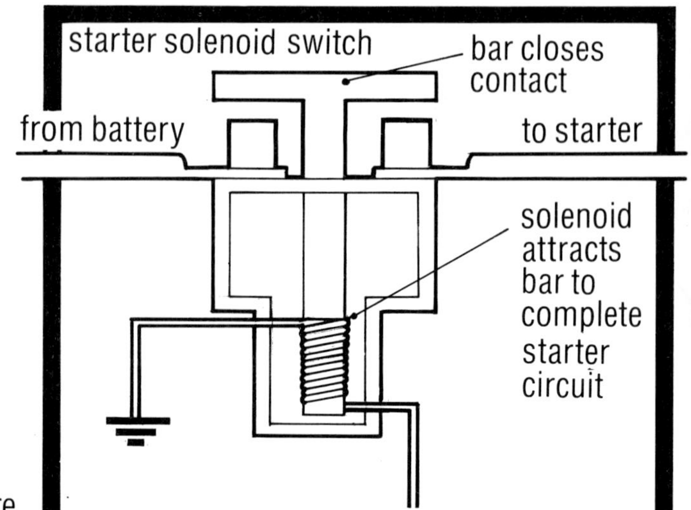

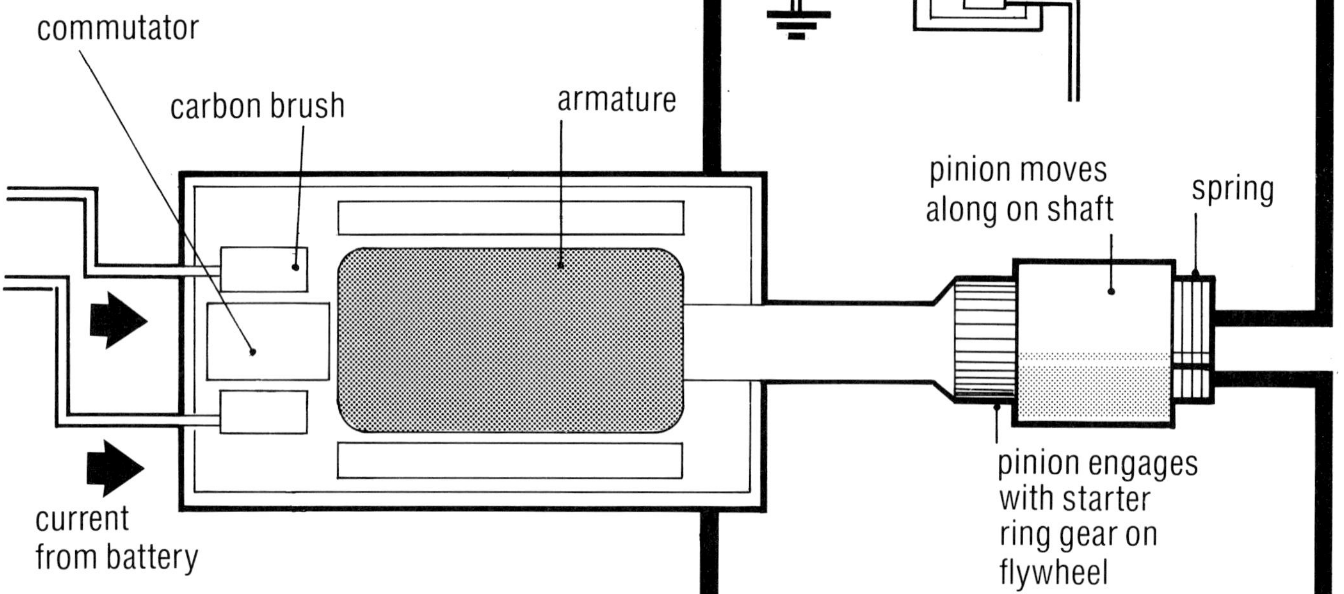

Starter solenoid switch

A thin wire from the ignition switch is connected to a coil of wire which forms an electro-magnet to pull a bar across contacts from the battery to the starter. This solenoid arrangement avoids the problems of running a long, thick wire inside the car where it would be a fire risk.

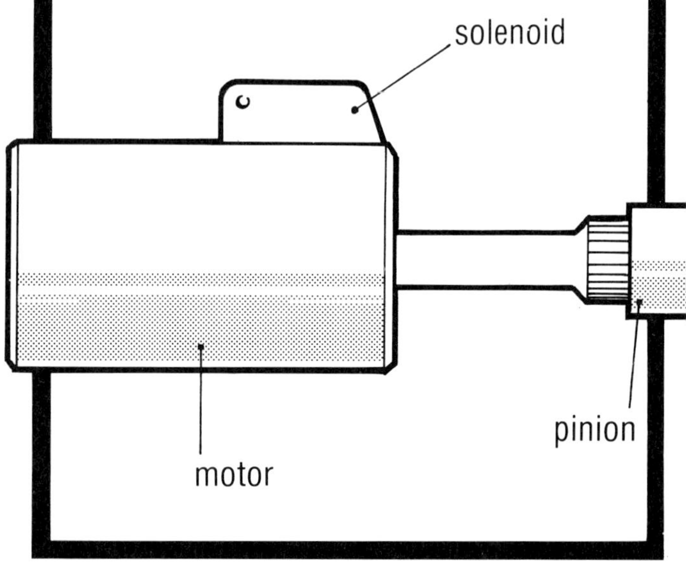

Pre-engaged starter

To reduce the wear on the pinion gear, pre-engaged starters move the pinion into mesh with the starter ring gear before the motor turns. A solenoid moves a lever which moves the pinion and then closes a contact to start the motor. This reduces wear but adds to the initial cost.

Lights

Headlights have a main beam for maximum light and dipped beam for use in traffic. Most headlights have two filaments (one for main and one for dipped) in a *sealed beam unit* comprising filament, reflector and lens. *Quartz halogen* lights give better illumination but cost more. Older lights were *pre-focus* lights with a bulb which was cheap and easy to replace but less reliable.

Sealed beam unit filament

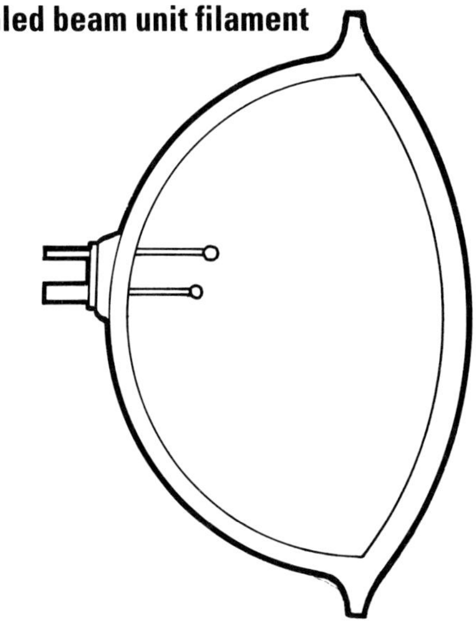

Twin filament bulb

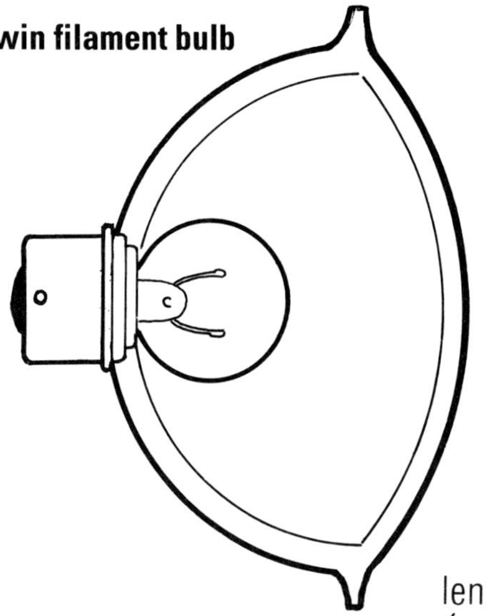

Rear lights use a bulb with two filaments, one for the normal light and one for the brake light. Indicator lights are switched on and off by a flasher unit.

Electrical terms

Voltage: Electrical pressure, the force which pushes electricity along (normally 12 volts).

Amperage: The rate of flow of electricity. A starter motor uses more electricity in a given time than does an indicator bulb.

Ampere hour: A measure of the capacity of a battery. A 10 amp hour battery will keep a 1 amp lamp lit for 10 hours or a 10 amp lamp lit for 1 hour etc.

Wattage: A measure of electrical power. A 100 watt bulb is brighter than a 60 watt bulb.

watts = volts × amps

EXAMPLE

2 rear lights, each 5 watts	= 10 watts
2 stop lights, each 20 watts	= 40 watts
1 number plate light, 2 watts	= 2 watts
Total power needed, 10 + 40 + 2	= 52 watts

$$\text{Current needed in amps} = \frac{\text{watts}}{\text{volts}} = \frac{52}{12}$$
$$= 4.3 \text{ amps}$$

The circuit carries 4.3 amps when everything is working properly. If a wire comes loose and touches the bodywork, causing a short circuit, a higher current will flow. Putting in a 5 amp fuse would safeguard the circuit.

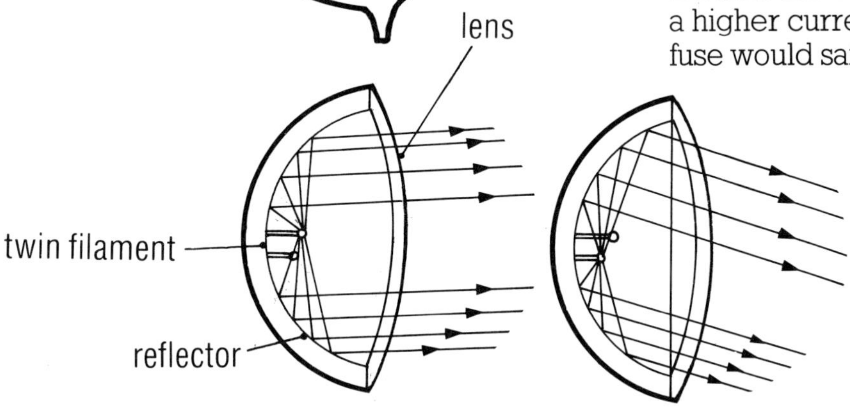

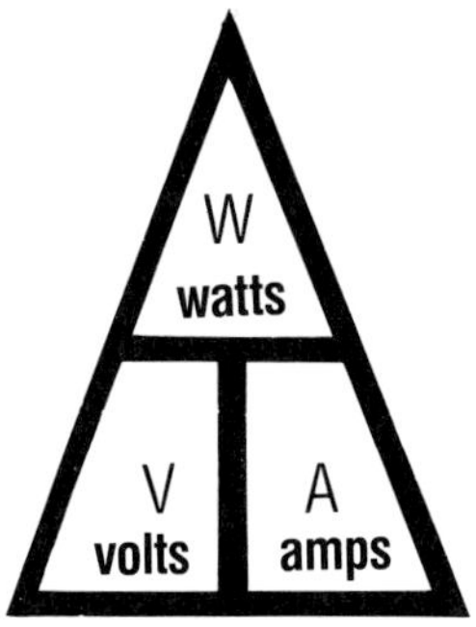

QUESTIONS

1 Write two or three sentences to explain the purpose of each of the following: starter motor, starter solenoid switch, main beam, dipped beam.

2 Calculate the fuse needed if the lighting circuit above were extended by adding two fog lamps of 40 watts each.

Practical: Change a sidelamp bulb and fit a replacement sealed beam unit. Remove, strip, clean, reassemble, test and replace a starter motor.

31. The battery

The battery is a store for electricity, made up of a series of *cells*. Each cell is basically two plates of lead in a solution of dilute sulphuric acid which produces an electric current of 2 volts. Six cells joined together will make a 12 volt battery.

The cell works by the sulphuric acid splitting in two parts to form positively charged hydrogen and negatively charged sulphate which move towards the positive and negative plates. This movement of 'electrons' is the basis of the flow of electricity.

The positive plate is lead peroxide (PbO_2) and the hydrogen combines with this and the sulphuric acid (H_2SO_4) to form lead sulphate ($PbSO_4$) and water. The negative plate is spongy lead (Pb) and the sulphate combines with this to form lead sulphate ($PbSO_4$). As both plates become covered with lead sulphate and the sulphuric acid becomes water the reaction slows down and eventually stops. Charging the battery reverses this process.

H_2SO_4 splits into $2H^+$ and SO_4^{2-}
$H_2 + PbO_2 + H_2SO_4$ forms $PbSO_4 + H_2O$
$SO_4 + Pb$ forms $PbSO_4$

Hydrometer

If the relative density of the electrolyte is checked with a hydrometer the state of charge of the battery can be measured. A fully charged battery has a relative density of 1.3. If the battery is discharged and the electrolyte is nearly all water the relative density will be nearer 1.0.

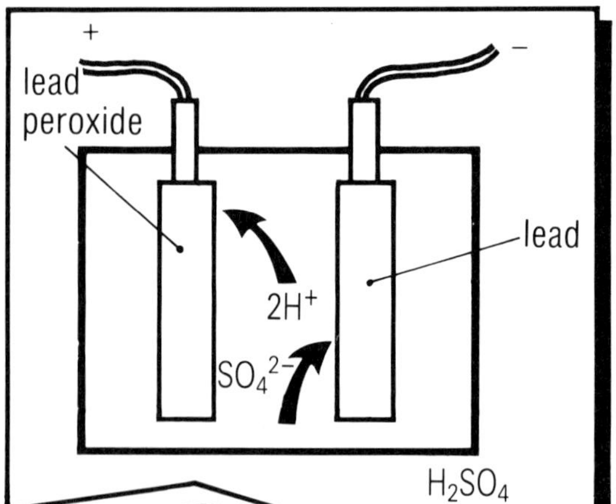

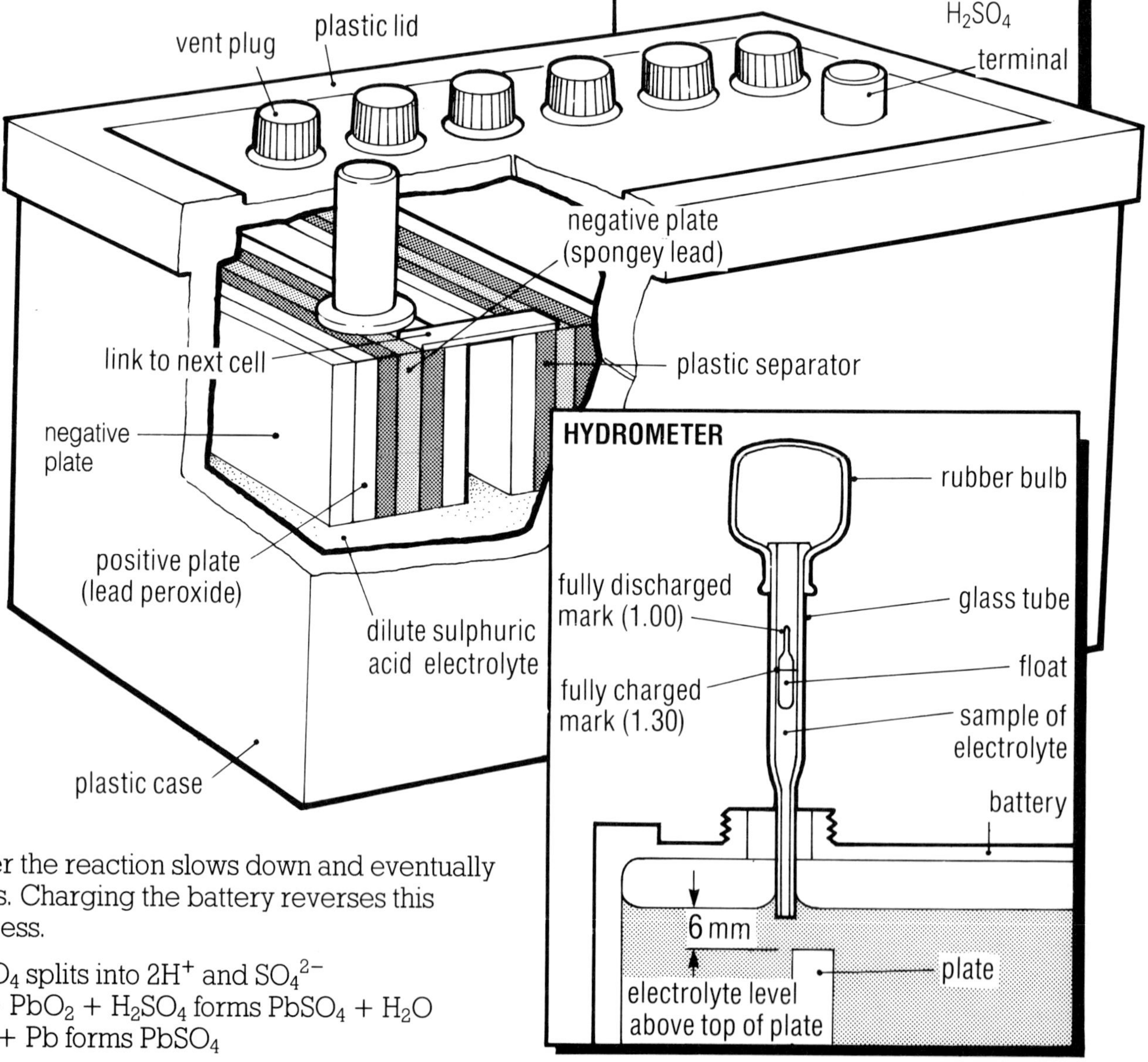

Battery charging

As electricity is used from the 'store' the battery becomes 'flat'. Charging the battery either from the dynamo or alternator or using a battery charger passes electricity back into it which pushes the sulphate from the plates into the water to reform sulphuric acid. The battery needs to be charged all the time that the engine is running and this is done by an electrical *generator*, either a dynamo or an alternator. Dynamos are simple and cheap but alternators produce more electricity at low speeds and are more efficient. Both are driven by a vee belt from the crankshaft pulley.

Alternators produce a great deal of current even at low engine speeds which is important in modern cars fitted with electric windows, radio, heated rear windows, etc.

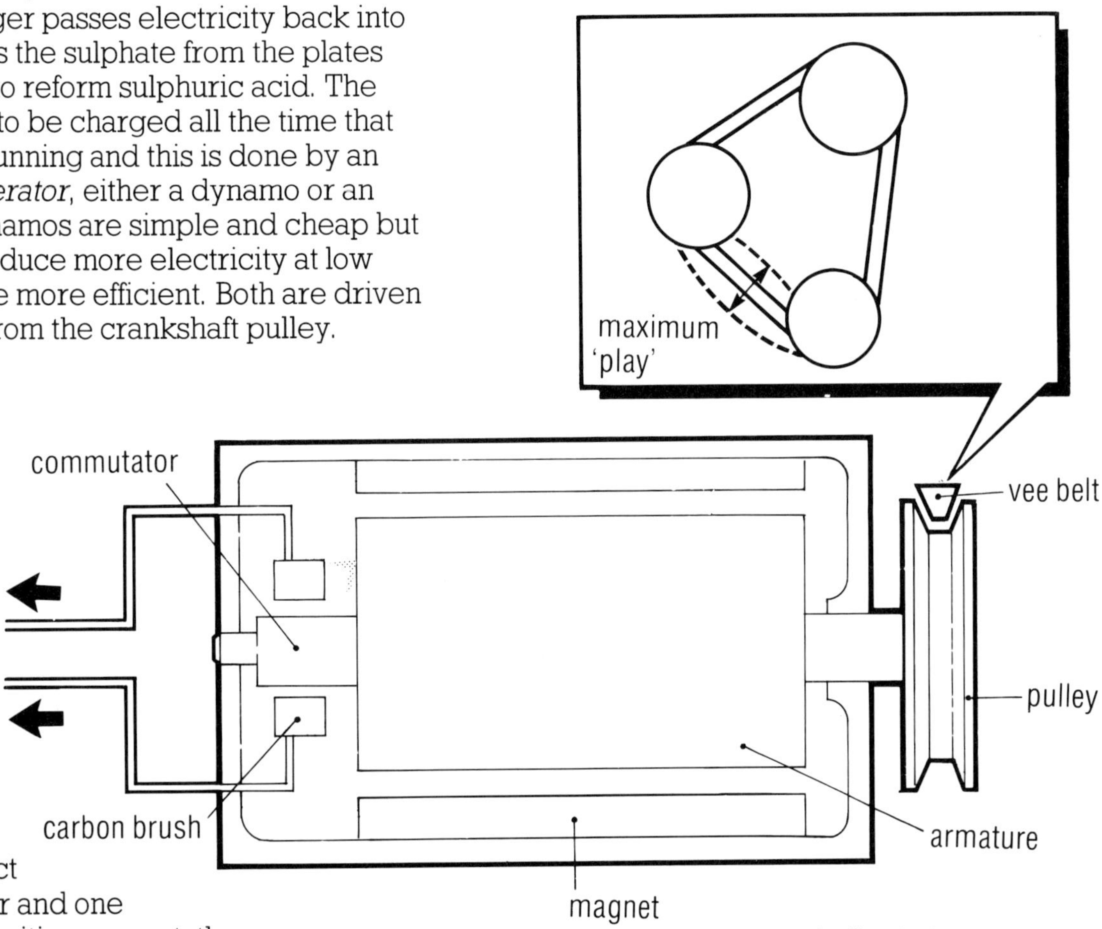

DYNAMO

An *armature* containing coils of wire is rotated between two magnets. This induces an electric current which is passed to a *commutator*. Two carbon *brushes* contact the commutator and one picks up the positive current, the other the negative current. The dynamo produces direct current (DC) the same as the battery and a voltage and current regulator controls the charge to the battery.

ALTERNATOR

A magnetic armature is rotated between coils of wire to induce an electric current. The alternating current (AC) which is produced is changed to direct current (DC) by a *rectifier* before it can be used to charge the battery.

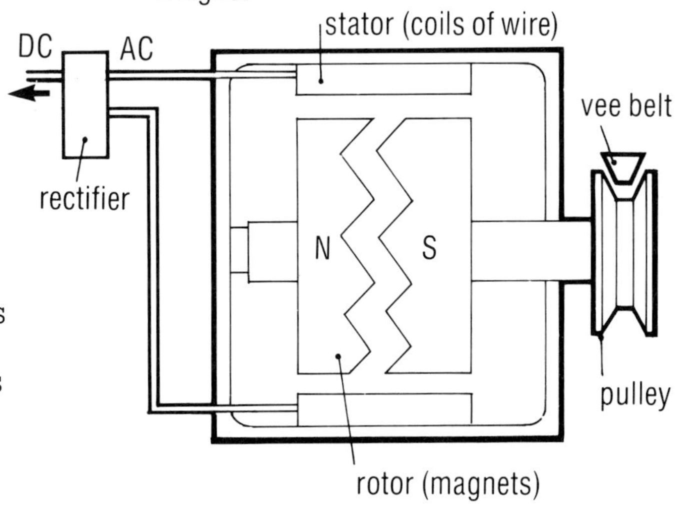

QUESTIONS

1 Explain, with sketches if necessary, how the battery works.

2 Make a neat, labelled drawing of a dynamo, explain how it works and why most modern cars have alternators fitted rather than dynamos.

Practical: Check the relative density of each cell of a 12 volt battery. Top up electrolyte to the correct level using distilled or de-ionised water.

EXAMINATION QUESTIONS

ELECTRICAL

1 What is the purpose of a fuse?

2 Name two materials which are electrical conductors and two which are insulators.

3 A battery needs 'topping up' with distilled water because
a) only the water evaporates or is lost by chemical action;
b) extra acid is formed by chemical action;
c) an old battery needs a weaker acid solution;
d) only the water leaks away through the battery case. (*NEA*)

4 Into which of the following circuits is the alternator/generator fitted:
a) lighting; c) charging;
b) starting; d) ignition (*NEA*)

5 The alternator has replaced the dynamo in modern cars because
a) it produces higher voltage;
b) it charges the battery even when the engine is only ticking over;
c) it produces d.c. rather than a.c.;
d) it is cheaper to manufacture. (*LEA*)

6 a) Redraw the illustration shown below in the form of a circuit diagram using the correct symbols.
b) Is this circuit
(i) negative earth;
(ii) positive earth? (*NEA*)

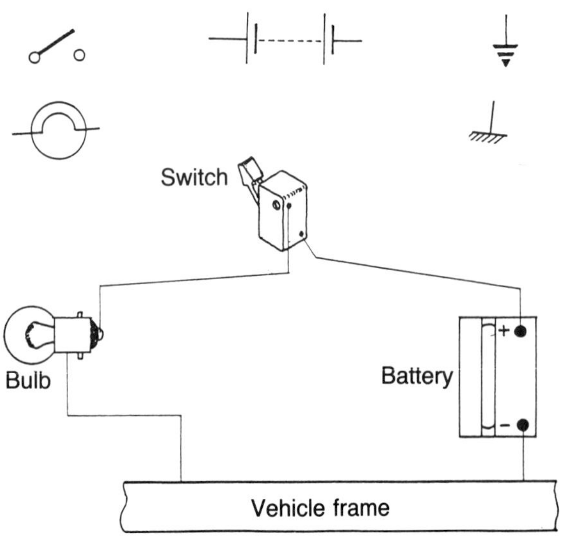

7 In the circuit below would failure of component *P* cause *Q* and *R* to
a) cease functioning
b) continue to function as before
c) function less efficiently
d) sustain damage? (*NEA*)

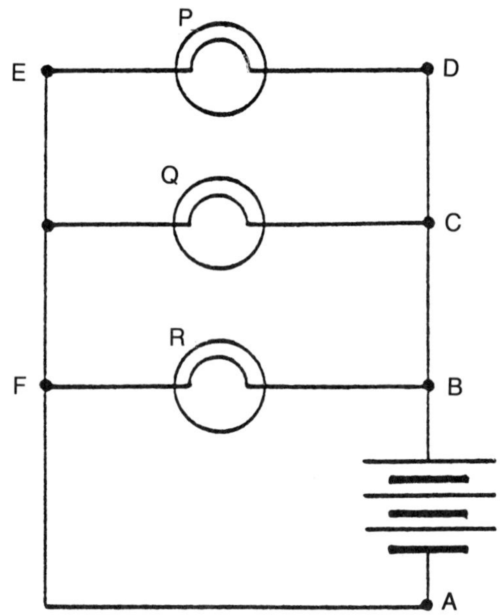

8 a) Draw a circuit diagram of a battery, switch, fuse and three lamps in series.
b) Draw a circuit diagram of a battery, switch, fuse and three lamps in parallel.

9 a) Make a neat, labelled sketch of a battery.
b) Explain how a battery works.
c) What is used to measure the relative density of a battery?
d) What is the relative density of a fully discharged battery?

10 a) What are the five main circuits in a car?
b) Which circuit uses red cable?
c) Calculate the fuse required for a circuit which comprises 2 rear lights each 5 watts, 2 stop lights each 20 watts and 2 rear fog lights each 40 watts.

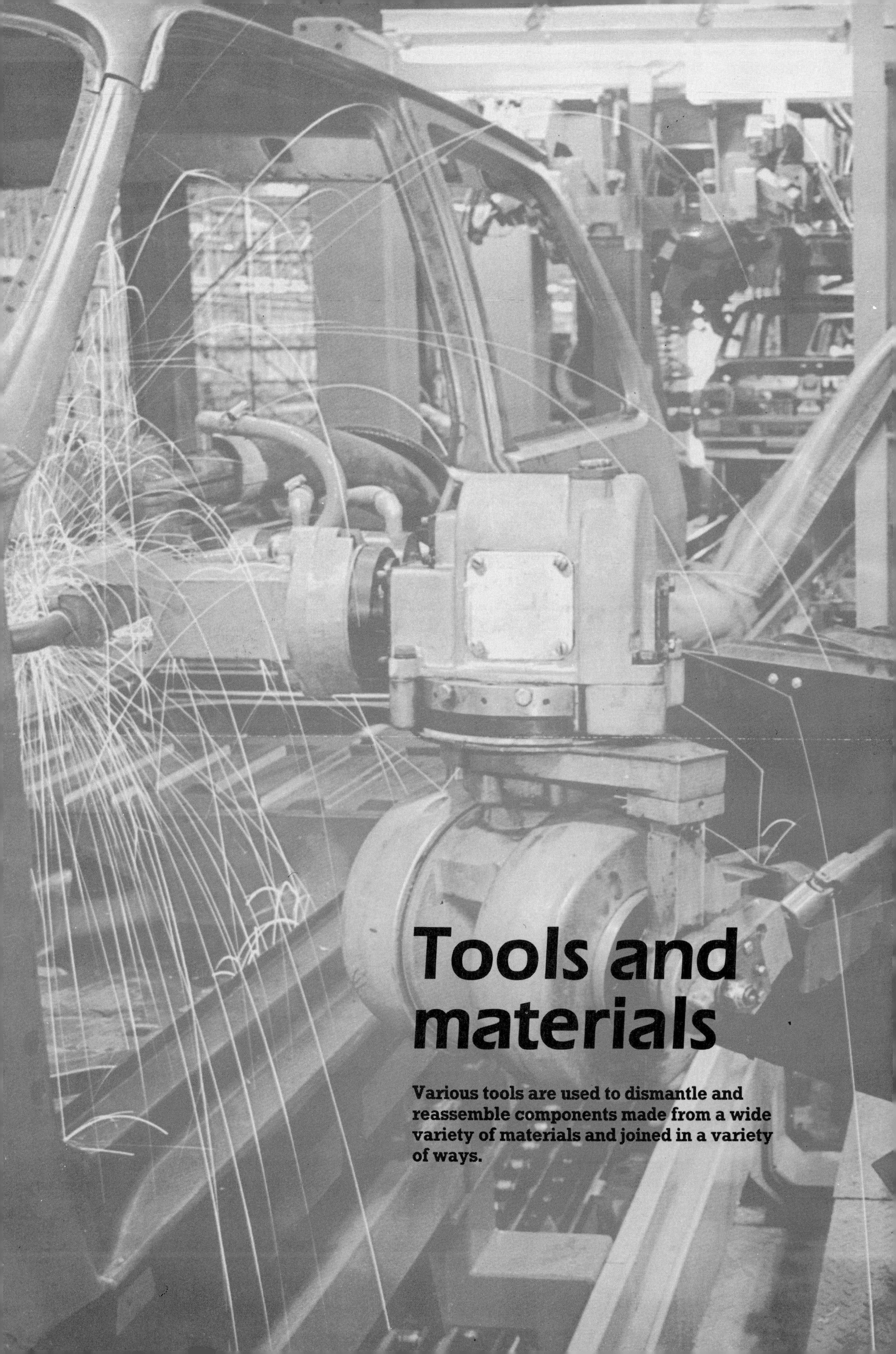

Tools and materials

Various tools are used to dismantle and reassemble components made from a wide variety of materials and joined in a variety of ways.

32. Tools

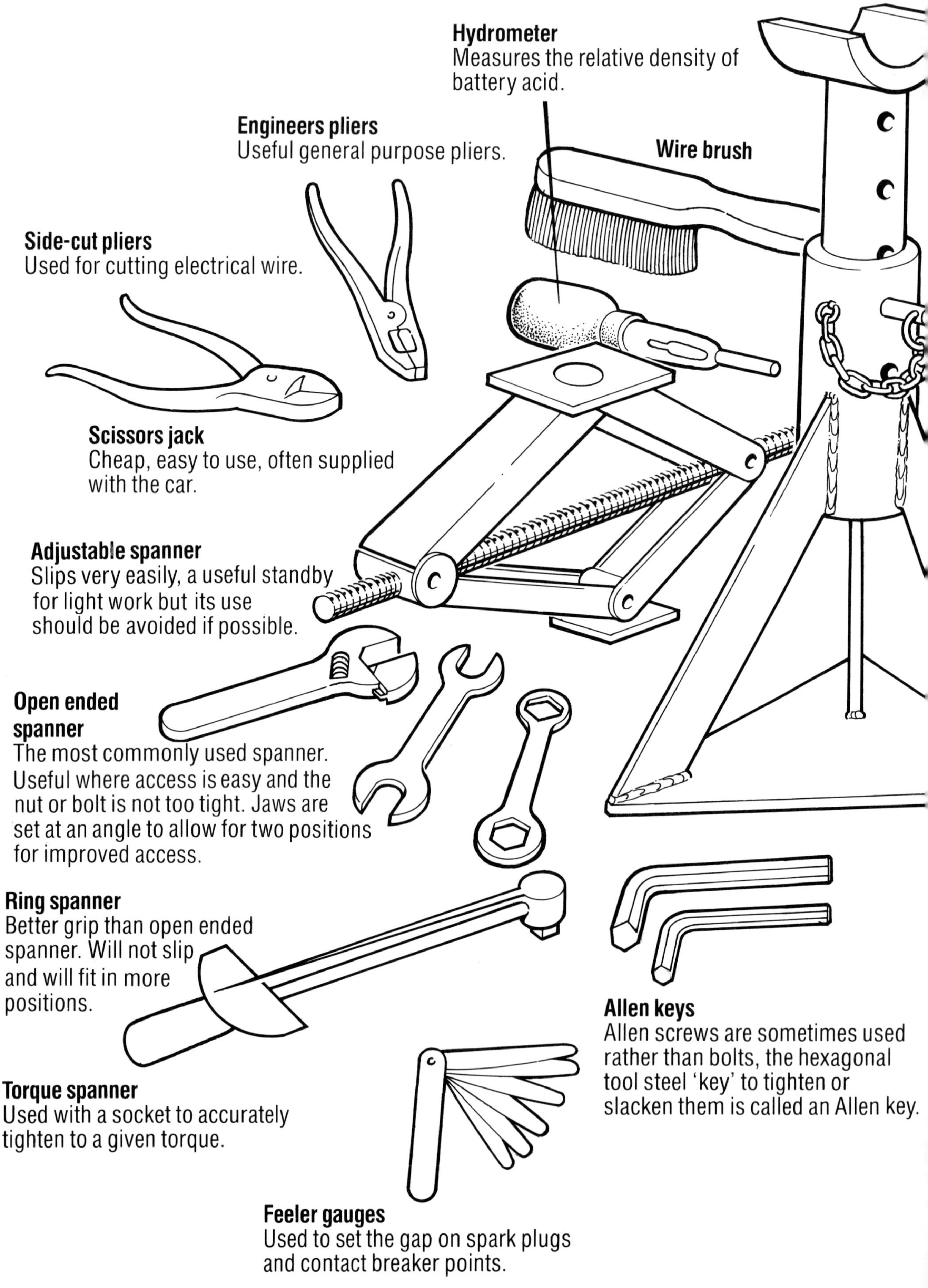
Hydrometer
Measures the relative density of battery acid.
Engineers pliers
Useful general purpose pliers.
Wire brush
Side-cut pliers
Used for cutting electrical wire.
Scissors jack
Cheap, easy to use, often supplied with the car.
Adjustable spanner
Slips very easily, a useful standby for light work but its use should be avoided if possible.
Open ended spanner
The most commonly used spanner. Useful where access is easy and the nut or bolt is not too tight. Jaws are set at an angle to allow for two positions for improved access.
Ring spanner
Better grip than open ended spanner. Will not slip and will fit in more positions.
Torque spanner
Used with a socket to accurately tighten to a given torque.
Allen keys
Allen screws are sometimes used rather than bolts, the hexagonal tool steel 'key' to tighten or slacken them is called an Allen key.
Feeler gauges
Used to set the gap on spark plugs and contact breaker points.

xle stands
sed to support a car. Must be ɔsitioned on a level surface and ıder solid part of the car.

Files
Various types of files are used the most common being rough, second cut and smooth, in a variety of shapes.

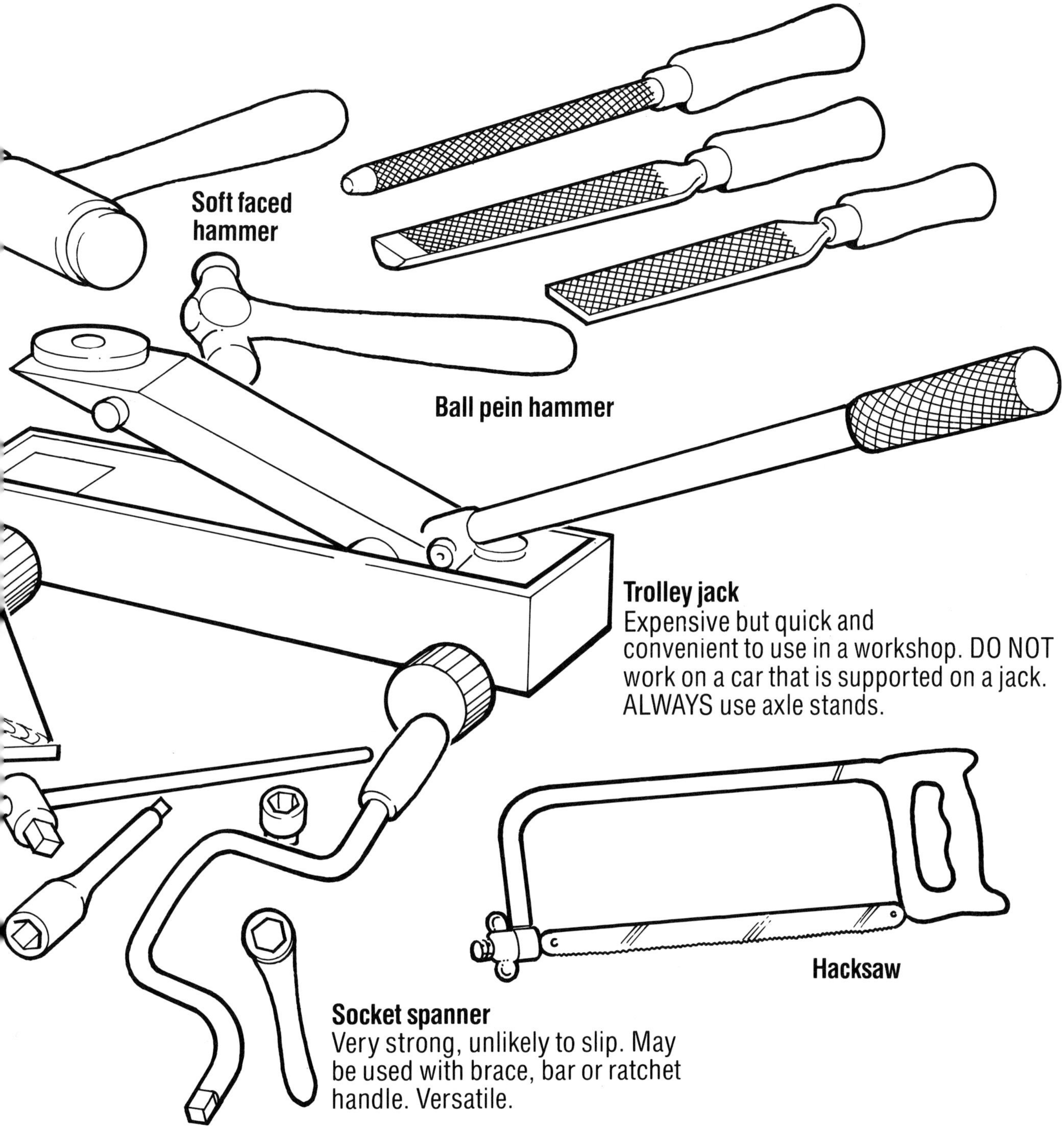

Trolley jack
Expensive but quick and convenient to use in a workshop. DO NOT work on a car that is supported on a jack. ALWAYS use axle stands.

Socket spanner
Very strong, unlikely to slip. May be used with brace, bar or ratchet handle. Versatile.

QUESTIONS

1 Make a list of the tools that you consider should be available in your own garage/ workshop at home.

2 Design a storage rack for the tools that are used in a workshop. The rack should protect the tools and be convenient to use. Add notes to explain the important design features.

33. Materials

Many different materials are used in the manufacture of a car. The diagram shows some of the more common ones.

The choice of which material to use depends on:

- Cost
- Properties
- Ease of manufacture

Often the choice is between what is wanted and what can be afforded. Stainless steel exhaust systems last longer than mild steel but cost more. Mild steel body panels rust but are strong, easy to produce and reasonably cheap. Aluminium cylinder blocks are lighter but more expensive than cast iron.

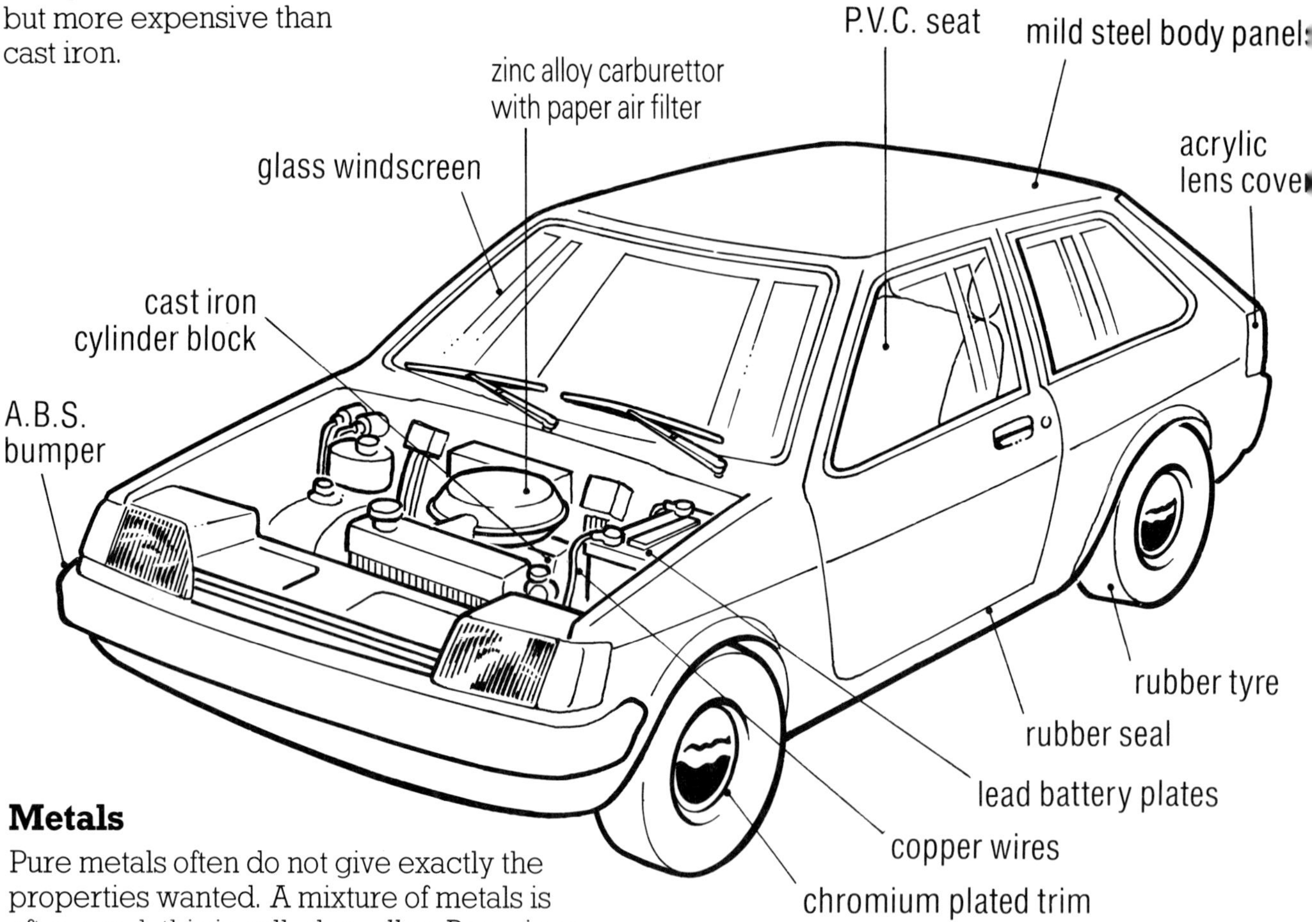

Metals

Pure metals often do not give exactly the properties wanted. A mixture of metals is often used, this is called an *alloy*. Brass is an alloy of copper and zinc, bronze is an alloy of copper and tin. *Non-ferrous* metals contain no iron; *ferrous* metals all contain some iron (see charts below). Steel is an alloy of iron and carbon; the amount of carbon affects the hardness of the steel.

Plastics

There are two main types of plastics:

Thermosetting plastics are hardened and set by heat when they are initially formed. Reheating melts them.

Material	**Properties**	**Uses**
Copper	conducts electricity, bends without cracking	electric cables
Brass	copper/zinc, strong, no rust	radiators
Bronze	copper/tin, strong bearing material	plain bearings
Aluminium	light, strong, no rust	bodywork
Duralumin	aluminium/copper/manganese, light, strong	pistons
Y alloy	aluminium/copper/nickel, strong at high temperatures	pistons
Zinc	easy to cast accurately	die castings
Lead	soft, alloyed to make 'white metal'	battery, big end shell bearing

Material	**Properties**	**Uses**
Mild steel	cheap, easy to shape	body panels, brackets etc
Medium carbon steel	medium strength, cheap	nuts, bolts and washers etc
High carbon steel	hard but brittle	gears
Cast iron	hard wearing, easy to cast	cylinder blocks
Malleable cast iron	hard and tough	connecting rods
Manganese steel	tough	springs and axle
Nickel steel	strong and tough	axles, suspensio arms
Chromium steel	does not rust, looks good	trim, bumpers, exhausts

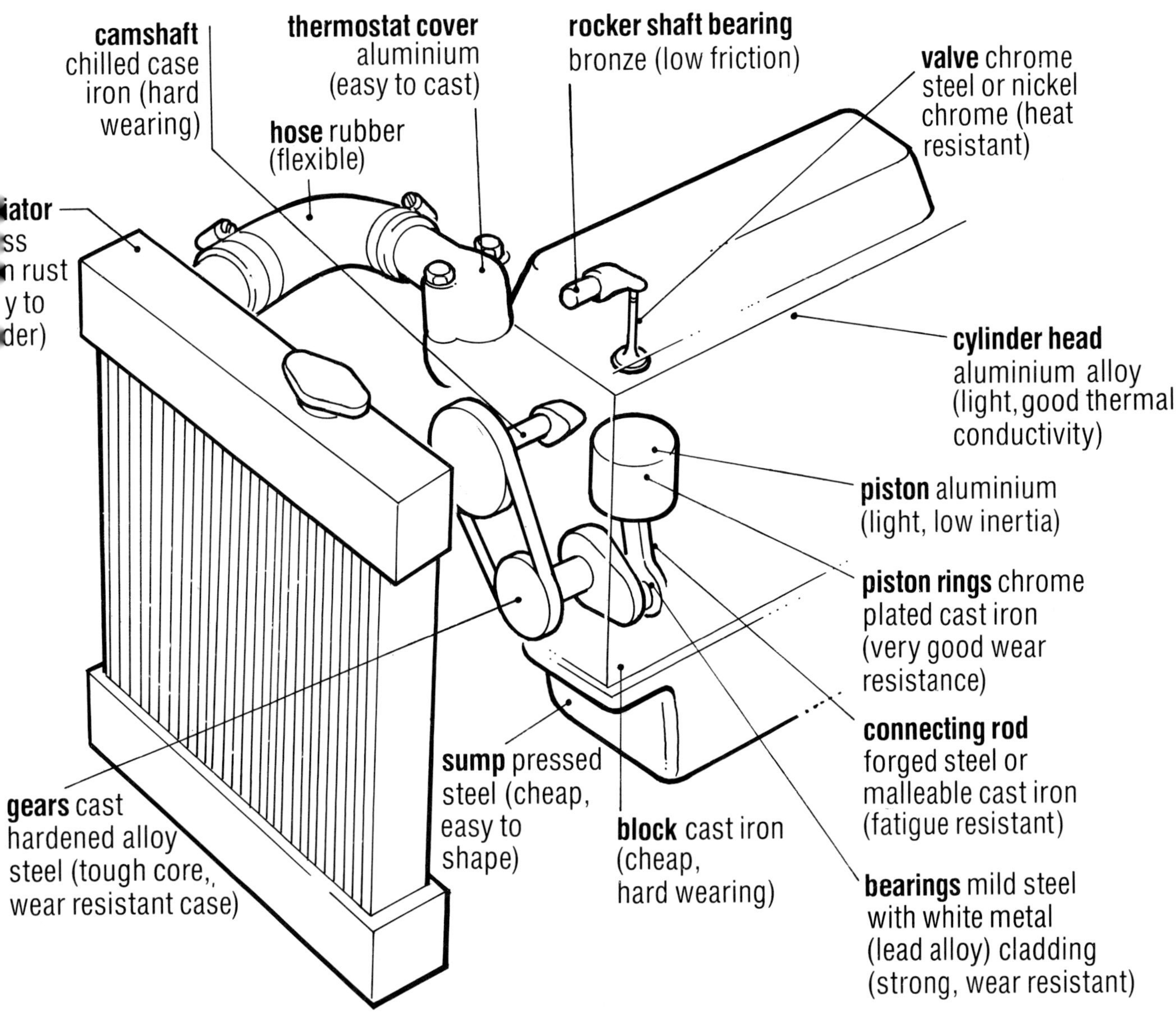

Thermoplastic plastics can be reheated, reshaped and, when cool, stay in this new shape.

Plastics do not rot or rust and can be moulded into complex shapes. GRP bodies do not rust but are brittle and difficult to mass produce; they tend to be used in small production runs.

Plastic	**Common name**	**Properties/uses**
Polyester resin	GRP	non rust used for bodywork
Polyurethane		foam seats
Polyvinylchloride	PVC	seat covers
Polymethylmethacrylate	Acrylic	Perspex lens covers
Polytetrafluourethylene	PTFE	non stick bearings and gears
Polyamide	Nylon	gears, light, rigid low friction
Amylbutylstyrene	ABS	shock absorbing bumpers, dashboard

This table gives some indication of the approximate weights of materials in a car which weighs 1000 kg.

Ferrous metals	800	Zinc	15
Glass	25	Copper	15
Rubber	25	Lead	10
Plastics	25	Others	60
Aluminium	25		

QUESTIONS

1 Draw a neat diagram of a car and label the various materials used in its manufacture.

2 Draw a similar diagram to show the materials used in an engine.

34. Joining

Permanent joints

SOLDERING

Solder is an alloy of lead and tin. The metal surfaces to be joined are cleaned and the solder melted and allowed to flow between them. When the solder cools it forms a solid joint. Flux is used to keep the surfaces clean and help the solder to flow.

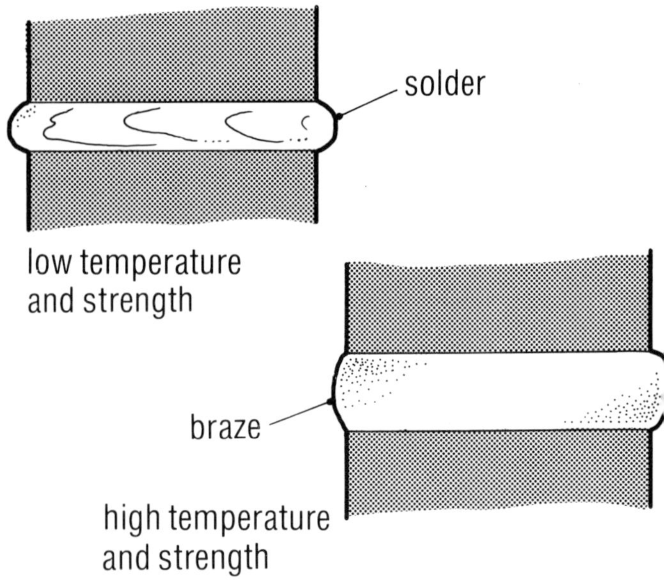

BRAZING

Similar to soldering but using brass to form the joint. A higher temperature process but gives a stronger joint than soldering.

WELDING

Oxygen and acetylene gas (oxy-acetylene) or electric spark (arc) is used to melt the metal. When spot welding, electrodes are placed either side of the metals to be joined and the resistance to the flow of electricity causes heat which melts the metal. This gives a very strong joint.

Fig 34 Arc welding wearing protective clothing

ADHESIVES

Various types of 'glue' have been developed to form permanent joints between all types of materials. They are quick and easy to use and form relatively strong joints.

RIVETS

The joint is made by putting rivet through holes drilled in the materials to be joined and forming a second 'head' to hold them together. There are various types of head such as *countersink* and *snap*. Pop rivets may be used in a confined space.

Temporary joints

NUTS AND BOLTS

The bolt passes through clearance holes drilled in both parts to be joined. Used where there is space to use spanners. Older British cars used a thread called British Standard Whitworth (BSW or Whit.) or British Standard Fine (BSF). More modern cars use a Unified thread, either Unified Fine (UNF) or Unified Coarse (UNC) and the spanner size is measured as the distance 'across the flats' of the hexagon (AF). On the continent metric threads are used.

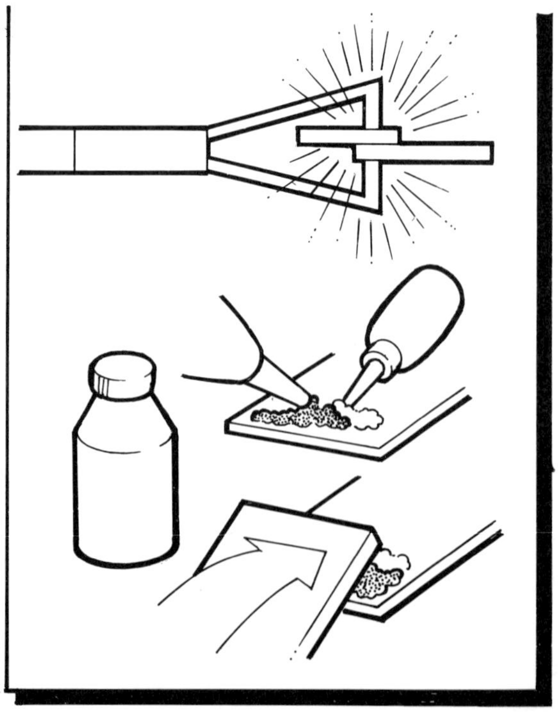

STUDS AND NUTS
Only one spanner is needed if a stud is screwed into a large component.

SELF-TAPPING SCREWS
Used when fixing light components together; quick and easy to use.

LOCKING DEVICES

Castle nut: Used with a split pin to give a positive mechanical lock.

Simmonds nut: Slot cut into the thread provides firm grip.

Nylock nut: Nylon insert grips onto metal thread and prevents movement.

Lock nut: Two nuts tightened against each other will not shake loose.

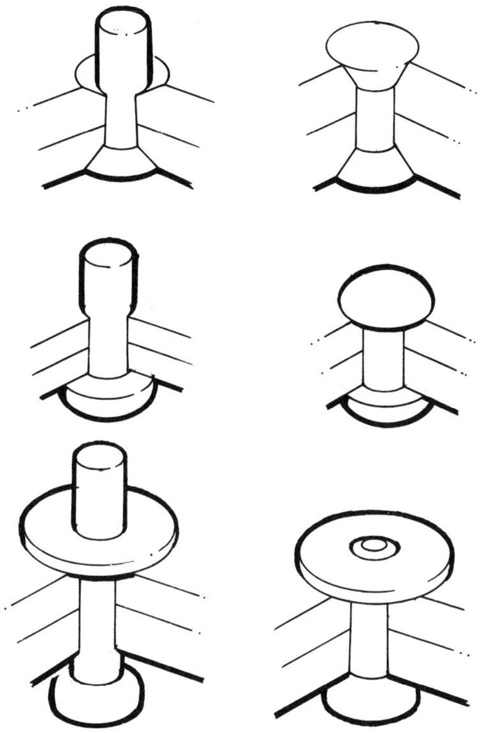

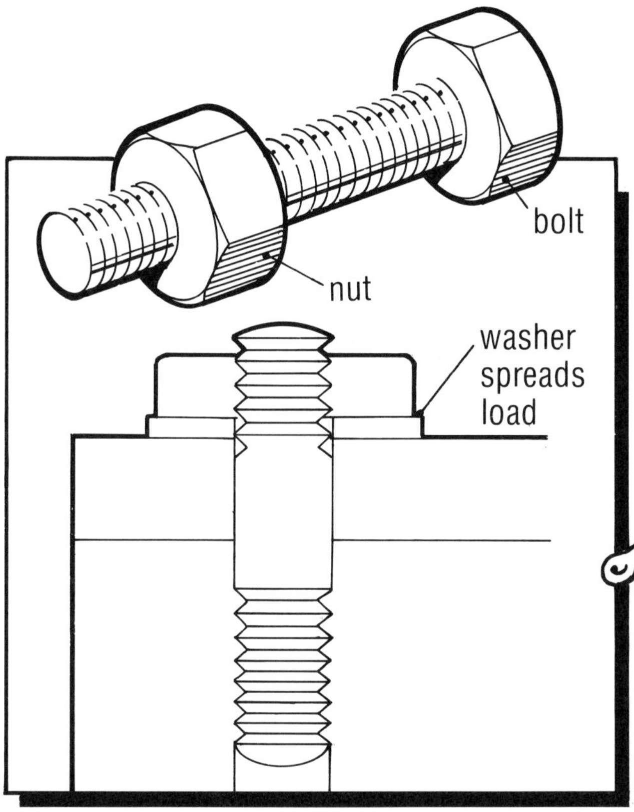

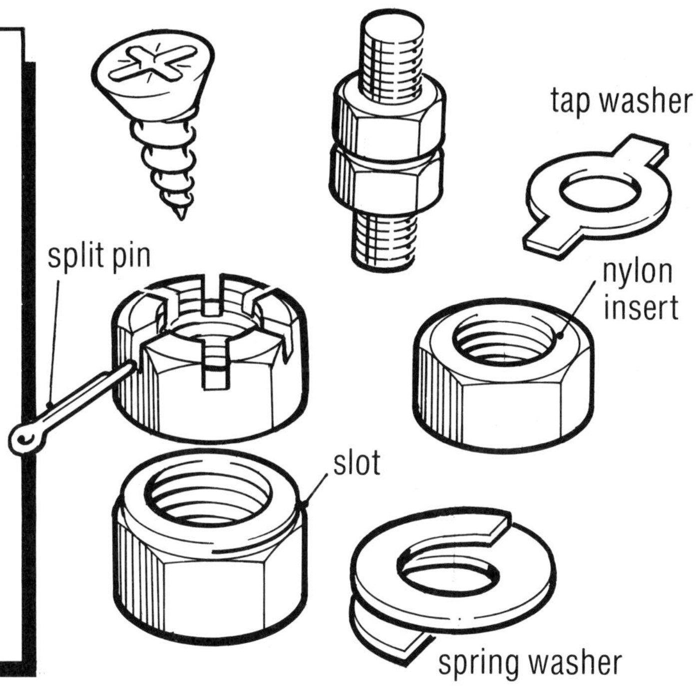

Spring washer: Prevents nut working loose.

Tab washer: Provides a positive mechanical lock.

Locking solution: A 'glue' that is put on the thread.

QUESTIONS

1 Make a list of the common types of fixing and give an example of where each might be used.

2 Draw neat sketches of each of the main types of locking devices and give an example of where each might be used.

EXAMINATION QUESTIONS

TOOLS AND MATERIALS

1 What is the advantage of a ring spanner compared with an open-ended spanner?

2 Name a material which can be used to make exhaust systems which should last the life of the car. (*LEA*)

3 What is the difference between ferrous and non-ferrous metals?

4 What do the following initials mean:
a) BSW b) UNF c) AF

5 Identify the materials used in the manufacture of the parts indicated in this diagram of a car.

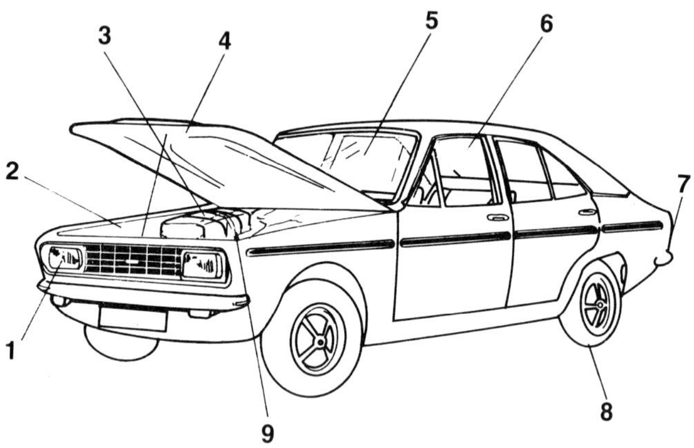

6 Give one example of a component made from each of the following materials and explain why this material is used:
a) copper
b) aluminium
c) PVC
d) mild steel
e) acrylic
f) glass

7 Calculate the maximum turning effort (Nm) exerted by a spanner 250 mm in length when a force of 10 N is applied. Show clearly how you obtained your answer. (*NEA*)

8 What jointing method would be used in each of the following situations:
a) spot lamp bracket to car body;
b) decorative trim to bodywork;
c) joining two electrical wires;
d) holding a replacement steel panel in place prior to welding;
e) fixing carpet in position inside car.

9 What is the difference between thermoplastic and thermosetting plastics?

10 Why is a torque spanner used?

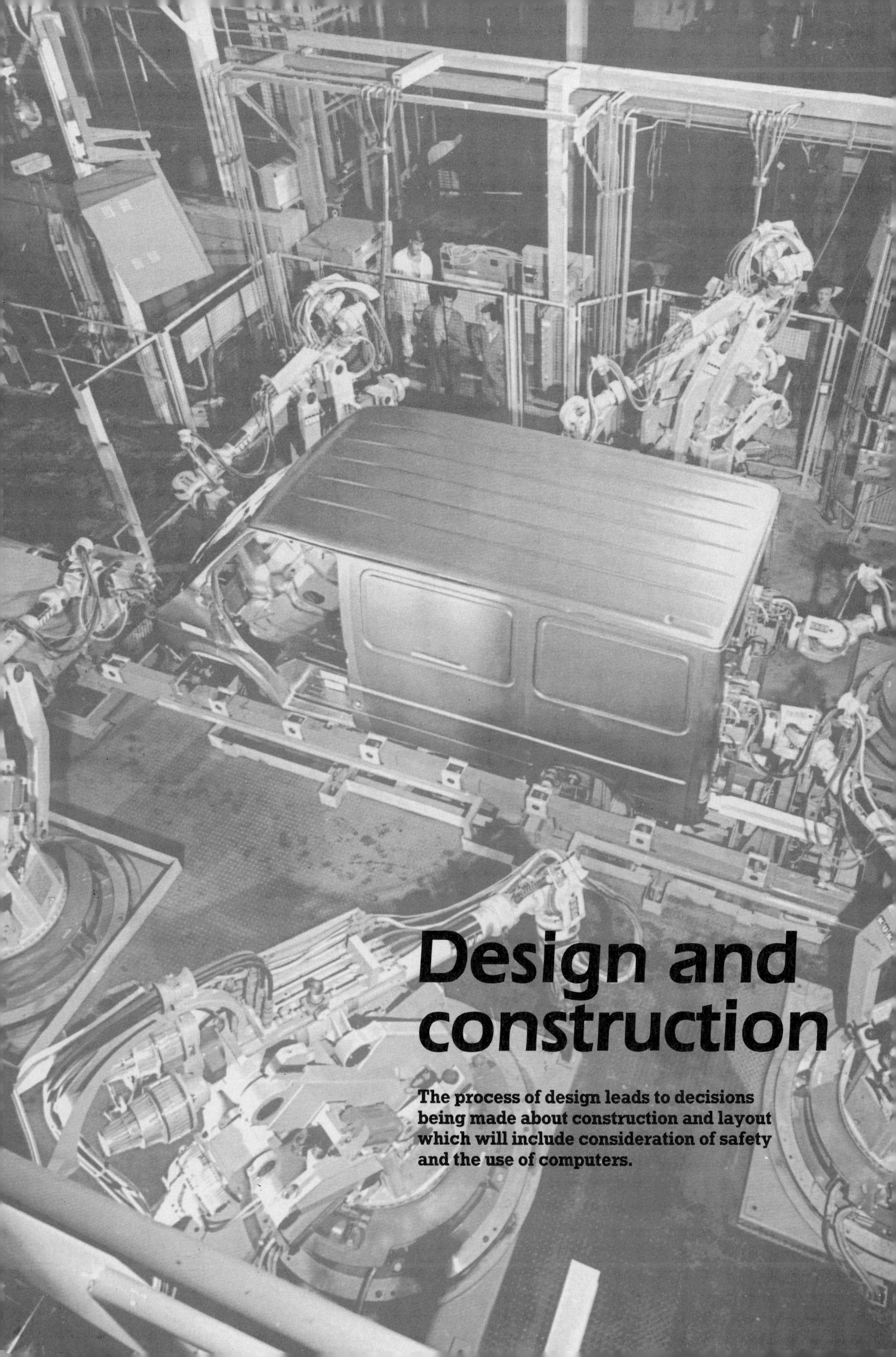

Design and construction

The process of design leads to decisions being made about construction and layout which will include consideration of safety and the use of computers.

35. Design

The design process consists of several stages:

1 *The problem:* What is needed?
2 *The design brief:* A written statement of the problem in detail.
3 *Information search:* Looking at other peoples ideas, notes and thinking about the problem.
4 *Alternative designs:* Several possible ideas to solve the problem.
5 *Selection:* Choosing the 'best' alternative.
6 *Implementation:* Developing the selected design, planning how it will be made, realising (making) it.
7 *Evaluation/testing:* Does it work? If not, why not and how can it be put right? If it does, can it be improved?

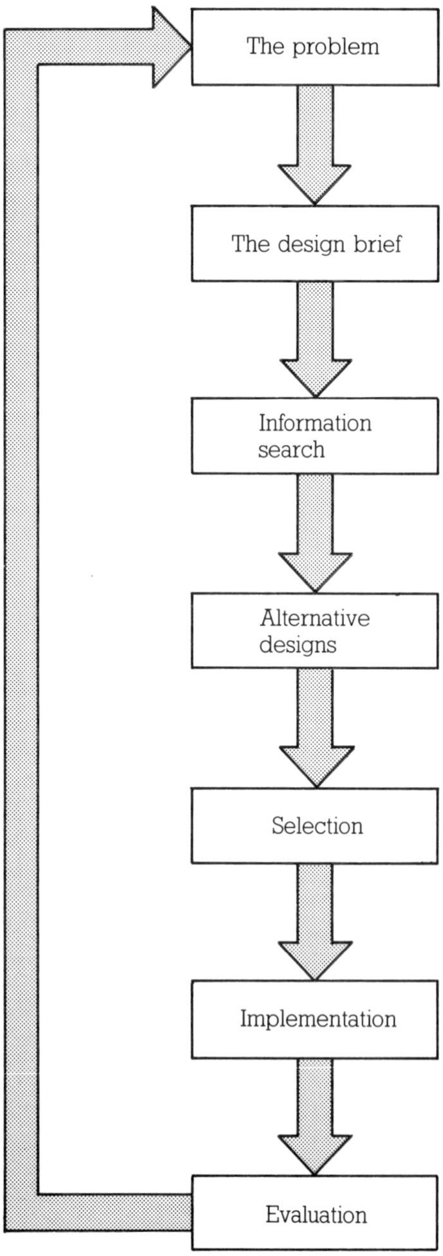

Design parameters

Car designers consider:

FUNCTION
What is it used for? Who will drive it? How many passengers? Long or short distances? Low or high speeds? Luggage capacity?

COMFORT
Ergonomics – comfortable and adjustable seating; controls and instruments in easy reach and sight; heating and ventilation.

ECONOMY
Cost of fuel, spares, servicing and insurance.

APPEARANCE
Style is very important and fashions are constantly changing.

PERFORMANCE
Speed, economy, acceleration and handling.

MAINTENANCE
Ease of fault finding and repair. Should a car be 'sealed for life' or accessible to the amateur?

SAFETY
See Chapter 37 on Safety in Cars.

MATERIALS AND CONSTRUCTION TECHNIQUES
Having been designed, the car it has to be made. A stainless steel car, for example, wouldn't rust but would be difficult to make, weigh a lot and cost the earth.

QUESTIONS

1 Make a list of the main stages in the design process.

2 What factors have to be considered in the design of a car?

36. Construction and layout

The design of cars developed from horse-drawn carriages with a short wheelbase chassis supporting an open high body.

CHASSIS TYPE
A strong steel frame supports the engine, transmission and body. Made from box section or channel section steel. Used in most cars until about 1940. Heavy, but body panels can be easily and cheaply changed. Used in some racing cars by small manufacturers (usually with glass fibre bodies) and in commercial vehicles.

SUB-FRAME TYPE
Front and rear parts of a chassis are bolted to the bodywork and support the engine and suspension units. Light, easier maintenance and quiet.

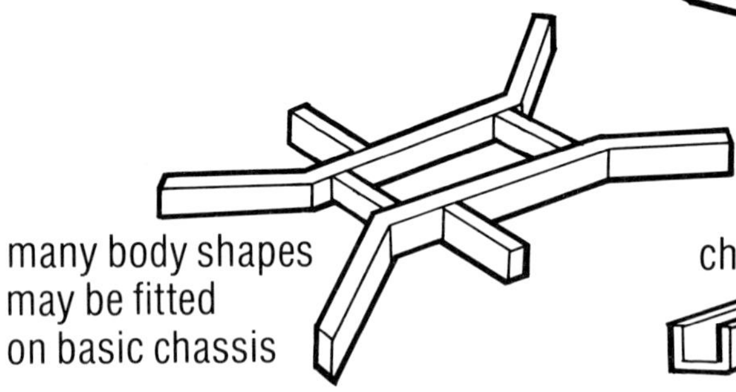

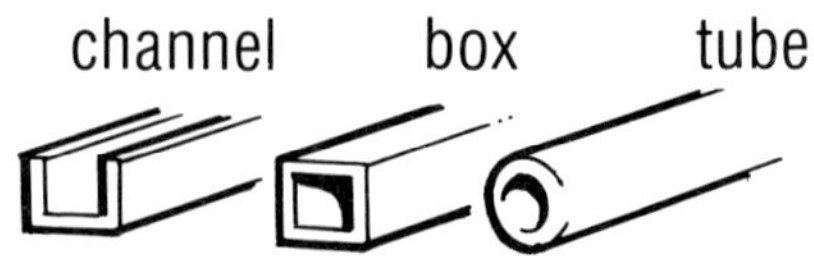

UNITARY TYPE
Aircraft type structure with the body acting as its own chassis. Steel panels welded together to form a shell which can take quite heavy loads. Strengtheners may be needed around sills, transmission, and bulkhead. Light, cheap to produce but expensive to repair accident damage.

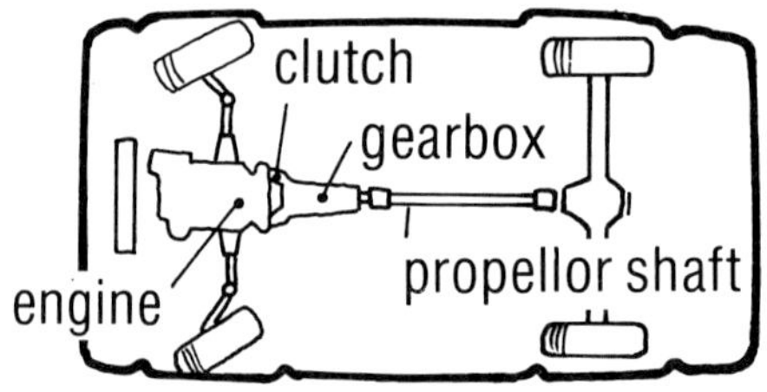

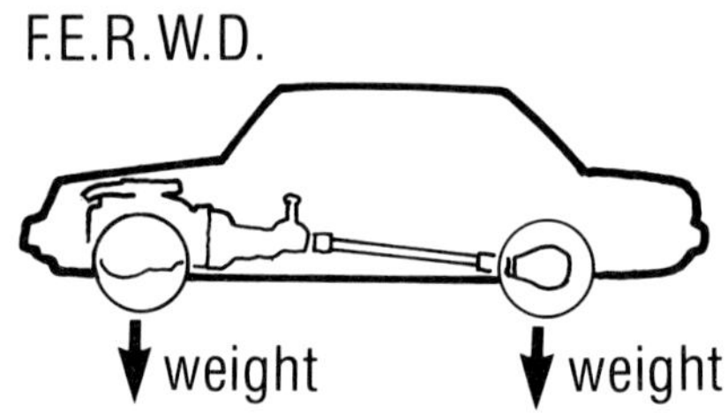

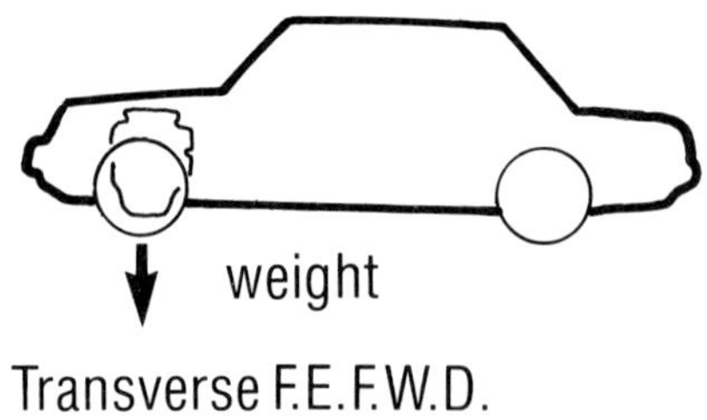

Layout
Front-engine rear-wheel drive:
The traditional layout, with each part of the transmission system separate and easy to work on. Weight distribution is good.

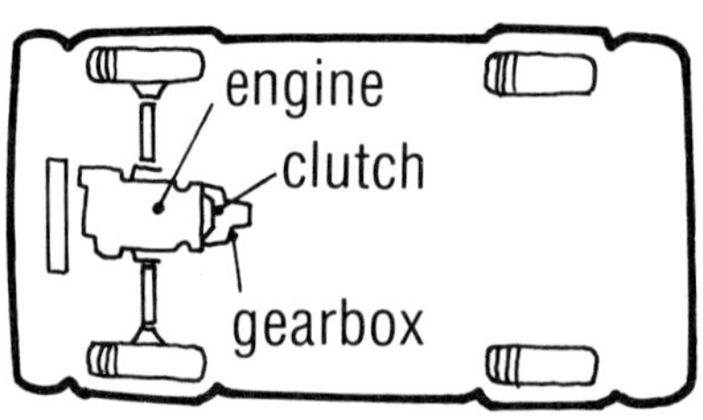

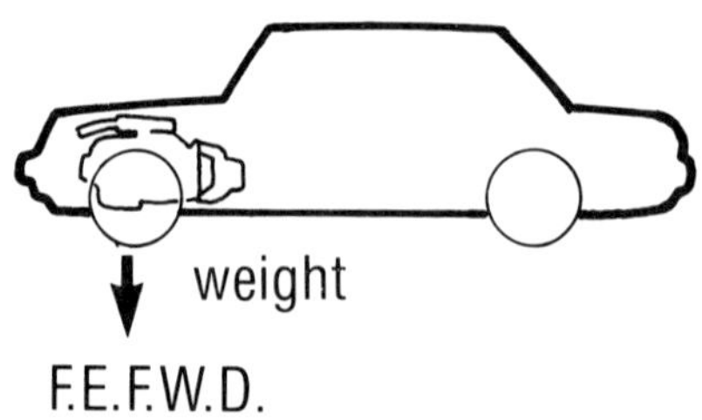

Rear-engine rear-wheel drive:
Similar advantages and disadvantages to front-engine front-wheel drive but simpler drive shafts. Weight distribution poor.

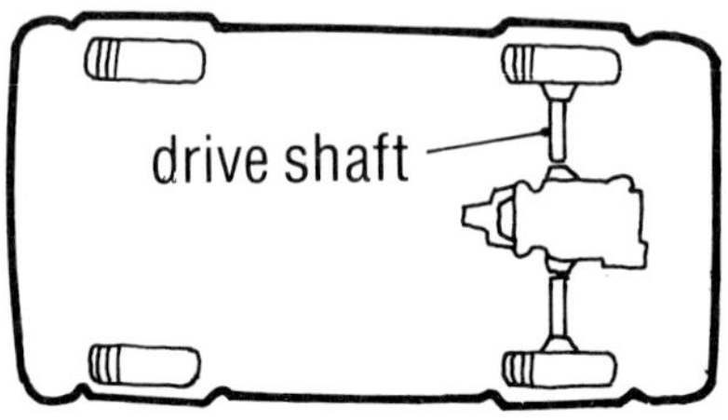

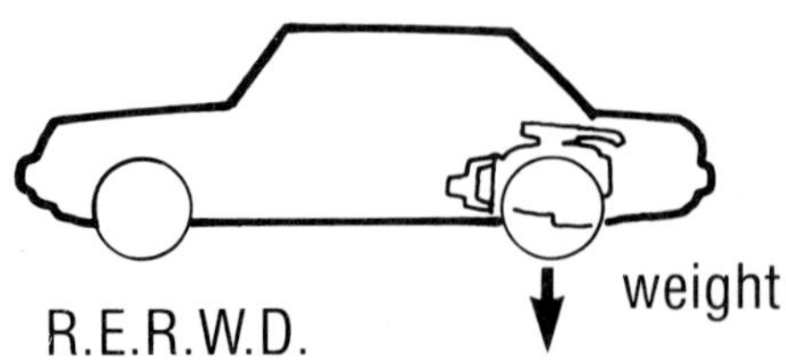

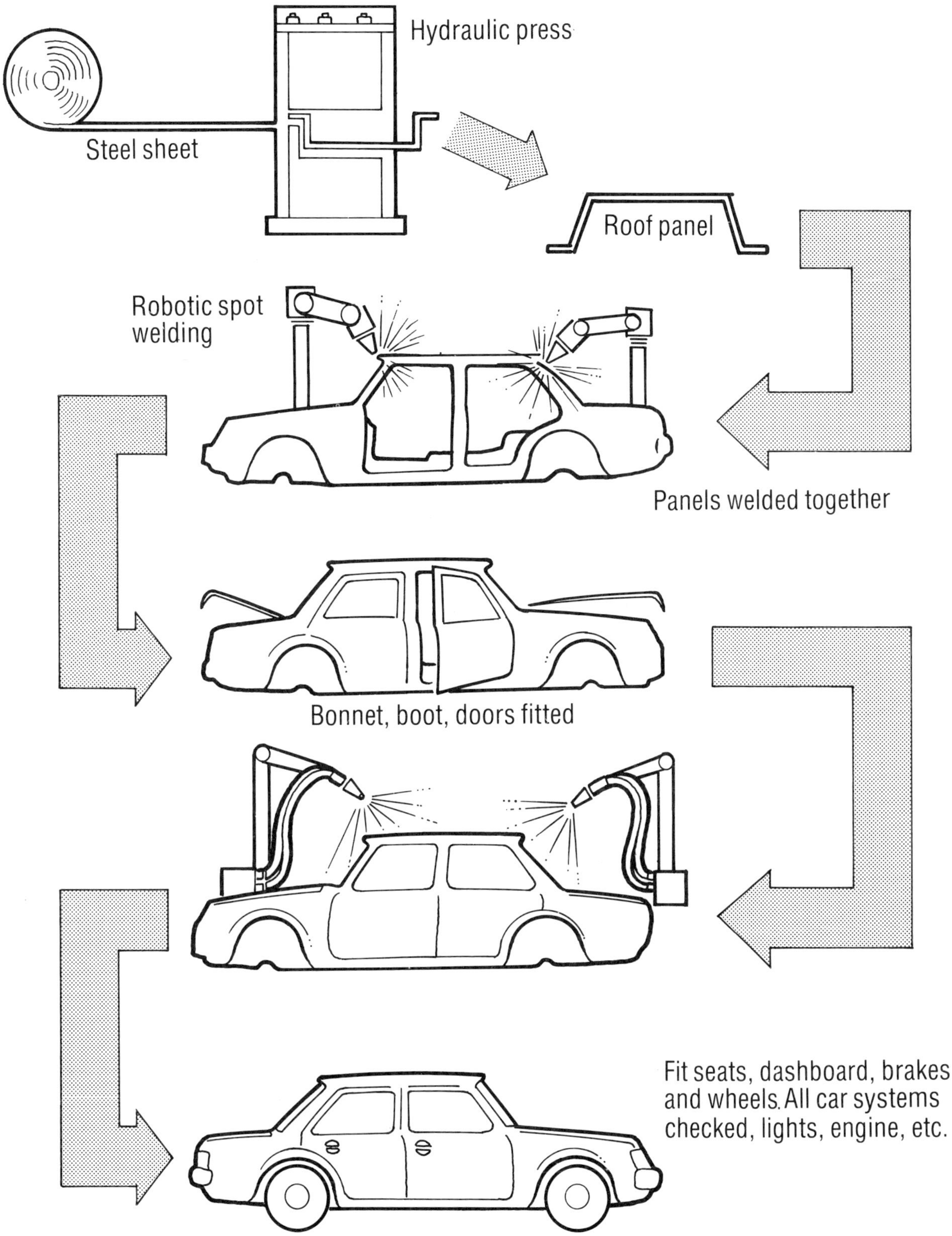

Front-engine front-wheel drive:
Weight over driven wheels gives good roadholding. More room inside passenger compartment because there is no transmission tunnel. More space created if engine is turned sideways (transverse engine) but this requires complicated drive shafts as driven wheels also have to steer.

QUESTIONS

1 Explain the advantages and disadvantages of chassis, sub-frame and unitary construction.
2 Write a short summary of the main stages in the manufacture of a car.

37. Safety in cars

There are two aspects to safety in cars:

1 Designing the car to reduce the risk of an accident happening.

2 Designing the car to protect the driver and passengers after an accident has occurred.

These two aspects are called *primary* and *secondary* safety.

Primary safety

1 Good visibility
2 Good layout of controls and instrumentation
3 Adjustable seating and steering
4 Dual circuit brakes
5 Good suspension for comfort and roadholding
6 Flexible engine for acceleration in any gear
7 Warning lights for brake failure etc
8 Good ventilation to prevent drowsiness
9 Child proof doorlocks
10 Effective lights

Secondary safety

1 Seat belts
2 Padded dashboard and headlining
3 Rigid safety cage with impact absorbing front and rear sections
4 Shock absorbing bumpers
5 Collapsible steering column
6 Safety glass windscreen
7 Burst proof doorlocks
8 Snap off mirror and switches
9 Flush mounted switches and handles
10 Head restraints to prevent 'whiplash'

Safety cage

Passengers are safe in a steel 'cage' which will not distort or crush easily. The bulkhead is particularly strengthened to prevent the engine being pushed into the 'cage' in an accident. The front and rear body sections are designed to collapse and absorb the energy of impact.

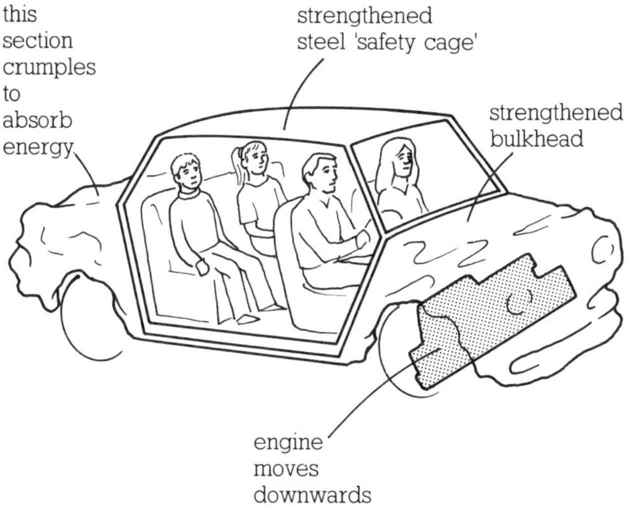

Department of Transport test

It is important that older cars are checked periodically for safety. The Department of Transport test is carried out on all cars three years old and older. If the car passes this test a certificate is issued which is valid for one year. It is illegal to use a car on the road if it does not have a test certificate unless it is under three years old. The test covers:

1 Brakes, which are tested on a rolling road.
2 Steering – there must not be excessive play in the ball joints etc.
3 Suspension, which affects the steering and so is very important.
4 Lights, which must work and be correctly aligned.
5 Tyres, which must have sufficient depth of tread and be in good condition.
6 Seat belts, which should be in good condition and the mountings not corroded.
7 Rust, as this affects steering, suspension, brake pipes and structural strength.

Whiplash injuries

In an accident the driver and passengers tend to get thrown forward towards the windscreen. But, having been thrown forward, the person's head then tends to 'flick' back and this can lead to a broken neck. Head restraints reduce the risk of these 'whiplash' injuries.

Drugs and alcohol

All drugs affect your control of your body. Sleeping tablets make you sleepy – they take control. If you are driving one tonne of metal at speeds of up to 70 mph it is vital that *you* are in control. Prescribed drugs normally carry warnings such as 'May cause drowsiness' or 'Do not drive'. Alcohol affects your judgement and your reaction time. Do not drink and drive. The penalty for driving while under the influence of drink or drugs is a heavy fine, usually loss of licence and even imprisonment.

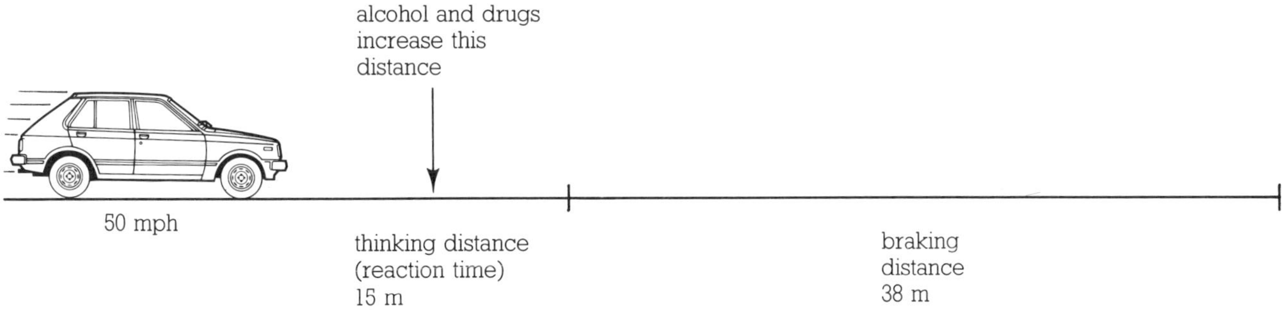

QUESTIONS

1 Explain, with examples, the difference between primary and secondary safety.

2 What items are covered in the Department of Transport test and how does it contribute to safety?

38. Commercial vehicles

Commercial vehicles are used for business purposes and range in size from the small van to the large specialised goods vehicles.

Smaller commercial vehicles

ESTATE CAR – often used by sales representatives and offering the alternatives of passenger carrying or goods capacity.

VAN – light goods vehicle with either hinged or sliding doors used for local deliveries.

PICKUP – open-backed vehicle often used by builders or decorators.

MINIBUS – small van conversion which can be driven by a driver with an ordinary driving licence.

Heavy commercial vehicles

Heavier loads need more wheels and more axles. An axle which is driven is called a 'live' axle, one which just carries the wheels and does not drive them is called a 'dead' axle. Small trucks are usually rigid, larger ones are *articulated* (a semi-trailer connected by a flexible coupling to a tractor unit). Engines may be fitted in the forward position or under the floor (mid-engined).

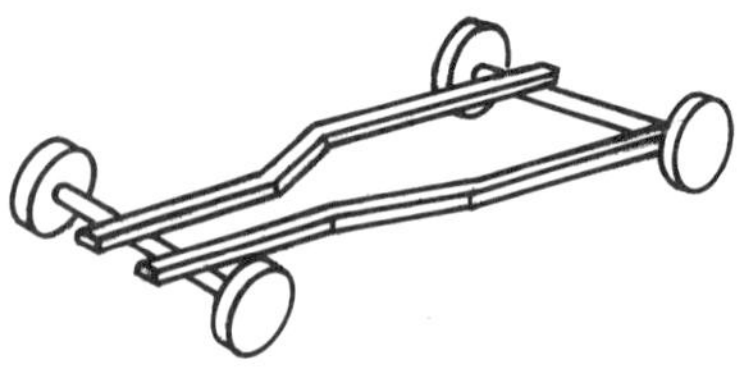

Bare chassis

Live axle

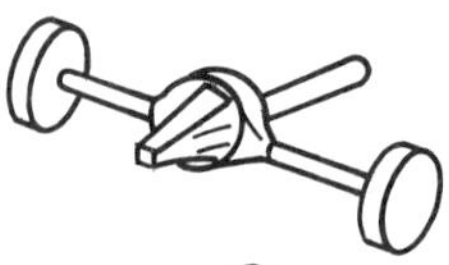

Dead axle

Twin wheels

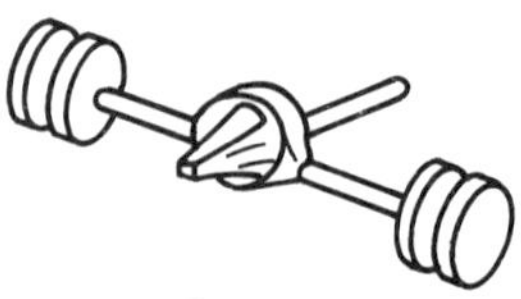

8 wheels

Fig 38(a) A heavy goods vehicle

Fig 38(b) A specialist body

Fig 38(c) An alternative body

Fig 38(d) A tractor unit and trailer

LORRIES are used for carrying heavy loads and usually powered by a diesel engine which offers long life, long service intervals and low fuel consumption. The bodywork is usually built onto a steel chassis so that one basic unit may have a range of alternative bodyshapes. Heavy lorries are Heavy Goods Vehicles (HGVs) and a special licence is required to drive them.

BUSES AND COACHES are used for carrying people and luggage. Long distance coaches have toilet, refreshment and sleeping facilities. A special PSV (Public Service Vehicle) licence is required to drive them.

TRANS-CONTINENTAL VEHICLES have to provide the driver with comfortable, self-contained accommodation and some are quite well appointed.

SPECIALISED VEHICLES are used for all manner of heavy transportation.

Fig 38(e) A public service vehicle

Fig 38(f) Pick-up body

Fig 38(g) An articulated tanker

Fig 38(h) BMC Nuffield 4/65 Tractor

QUESTIONS

1. Explain why commercial vehicles often have several axles.
2. Sketch the main types of smaller commercial vehicles.

39. Computers

Computers are used at all stages of the design and construction of cars, saving time and money as well as changing the types of work done by people employed in the industry. The first cars were handmade by craftsmen. Mass production techniques employed unskilled workers in simple, repetitive tasks reducing construction time and costs. Computers are taking over the unskilled elements of design and construction.

Computer aided design (CAD)

Designers use computers to examine their designs from different angles, on different scales and under different conditions. The ability to reproduce and alter drawings quickly, to perform stress calculations and send 'pictures' over the telephone have made the computer an invaluable tool for the designer.

Computer controlled machines

The designer's computer can produce programs to be used in machines which will produce components rapidly and accurately to a consistent standard. Production machinists are being replaced by machines which only require infrequent attention.

Robotics

Computer controlled robots are able to perform many complex assembly-line jobs. Repetitive tasks such as spot welding and painting, are particularly suited to robotics. Boring, monotonous work in unhealthy conditions, such as spray booths, are best left to computers.

Computer aided manufacture (CAM)

Cars are complex assemblies and the coordination of the work of many sub-assembly lines so that their components arrive on time at the main assembly line can be made far easier with computer assistance. Computer controlled stock transfer and computer controlled assembly of certain components has made vehicle manufacture more efficient.

Fig 39(a)(i) Human spot welding

Fig 39(a)(ii) Robotic spot welding

Monitoring and testing

Monitoring the quality of manufactured components is an area where the computer's ability to analyse test results and predict likely outcomes has made it invaluable.

Computers in cars

Computers are able to provide the driver with information on times, distances, fuel consumption and warn of mechanical defects, as well as controlling ignition and carburettion for optimum performance.

Fig 39(b)(i) Humans on a production line

Fig 39(b)(ii) Computer controlled production

Computers and jobs

The effect of computers has been that the boring, repetitive, unskilled and dangerous jobs are being given to machines whose speed, accuracy and low maintenance costs are an asset. This has inevitably meant fewer jobs for unskilled production workers.

Design, service, sales and so on still rely on human skills, ingenuity, adaptability and 'friendliness' (a computer would make a very poor salesman!). Some jobs may be replaced by computers, some may be made easier with the help of computers and some are best done by humans.

Additional, new jobs have been created in computer design, programming, installation and maintenance. These jobs are cleaner and more comfortable than the jobs which have been given to the machines.

Fig 39(c) CAD

QUESTIONS

1 Explain how computers have affected the design and manufacture of cars.
2 Write a short essay to explain how you think jobs have been affected by the introduction of computers.

40. Writing a project

Fig 40 Sources of Information

Two DO NOTs

1 DO NOT copy. You get no marks for copied notes. A photocopier is quicker and more effective at copying than you are.

2 DO NOT leave it to the last minute. A piece of work that was hastily scribbled down the night before will not gain many marks. You will be under less pressure if you take your time.

Getting ready

Think carefully before you start. If you have a choice, choose a topic that:

1 Will interest you

2 You will be able to get information on.

Plan what you are going to write about and how you are going to present your project.

Getting started

Write a one- or two-page introduction in your own words, something like:

'I have decided to do a project on . . . because my Uncle Phil works in a factory that makes . . .'

'Spark plugs are the most interesting part of a car and . . .'

'I think that . . . is really important and . . .'

'I know very little about . . . and this project may help me learn more . . .'

The introduction should explain what you are going to do and why. It should be short, written by *you* and make the person who reads it want to read on.

Finding information

Go to libraries, ask friends, visit garages and showrooms, write to manufacturers, buy magazines, above all *ask* for help. Read, try to make sense of what you have read, sort out the relevant information and write your first chapter in your own words.

You must make a list of all the sources of information that you use.

Diagrams

A project that is all writing is likely to be boring unless it is very well written. Most people prefer to add interest by using diagrams. Use: pictures, photos, cut outs, sketches, labelled diagrams, colour. Do not use: poor quality copies, pictures cut from expensive books, too many or irrelevant illustrations.

Remember that you are using illustrations for information and interest, you are *not* making a scrapbook.

How long?

It depends. A useful idea is to practice by doing a short (5 sides?) project, then a longer project before you attempt a main project. The important thing is the *structure* of the project – what is covered and in what order.

Make it interesting

You could use the odd cartoon, write about a visit to a factory, choose an interesting subject, make a model, include your own photographs, write up an interview with a mechanic, use overlay diagrams. Have a look at advertising brochures and see how they add interest. They are produced by professionals!

What the examiner looks for

Your project should show that you have researched, analysed and presented your project well. Examiners look at three main aspects:

RESEARCH
How well do you understand the subject? Is the information relevant? Have you collected information from several sources?

ANALYSIS
Has the information been analysed? Are any conclusions drawn? Are the conclusions valid?

PRESENTATION
Is the presentation good? Is it varied, clear and concise?

QUESTIONS

1 Make a list of the sources of information that you can use when writing a project.
2 Write an introduction for a project on the subject of spark plugs.

EXAMINATION QUESTIONS

DESIGN AND CONSTRUCTION

1 What are the stages in the design process?

2 Write a design brief for a small, economical car to be used mainly in town.

3 A motor car designed with the engine at the front which drives the rear wheels through a transmission system is called:
a) transverse;
b) open frame;
c) independent;
d) conventional. (*NEA*)

4 What are the main advantages and disadvantages of each of the following layouts:
a) front-engine rear-wheel drive
b) front-engine front-wheel drive
c) rear engine rear wheel drive

5 Explain the following terms
a) third party insurance,
b) comprehensive insurance,
c) no claims bonus.

6 Describe how computers are used in the motor industry and the effect that they have had on patterns of employment.

7 Although the accident rate is proportionately lower today than it was twenty years ago, there are still far too many people being killed or seriously injured. In your view, how could safety on our roads be improved? (*LEA*)

8 Using the letters *A* to *G* located against the list of vehicle components, complete each space located against the vehicle system in which the component is to be found.

The first has been completed as an example. (*NEA*)

Components	*Vehicle Systems*	
A Rack and pinion	1 Engine	G
B McPherson strut unit	2 Transmission	____
C Layshaft	3 Steering	____
D Sill	4 Suspension	____
E Solenoid	5 Body	____
F Master cylinder	6 Brakes	____
G Rocker cover	7 Electrical system	____

The use of specialist equipment and tools for repair and maintenance is an accepted and necessary part of motor vehicle engineering.

For a variety of reasons the equipment may not be available and it is left to the vehicle engineer to design and make suitable alternatives which will enable the job to be carried out safely and effectively.

Two problems are outlined below. You are asked to design a piece of equipment or tool which will satisfy each requirement and present your answer in the form of a sketch with appropriate explanations, including details of recommended materials.

9 This flywheel below and its ring gear must be stopped from rotating in either direction in order that pressure plate bolts *A* may be removed and refitted. (*NEA*)

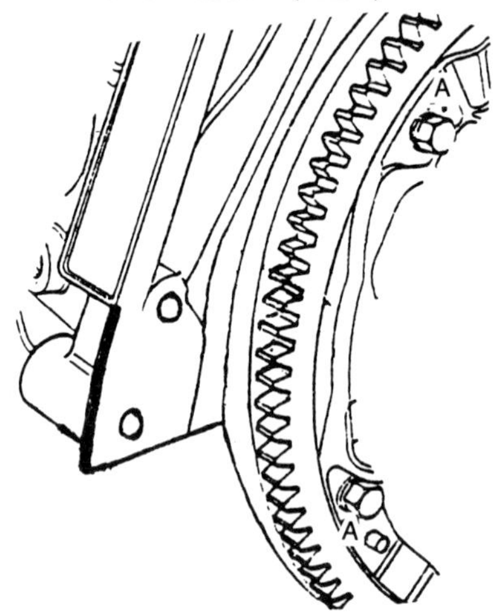

10 Valve Guide *D* must be pressed or driven out and a new guide fitted. This new guide must be accurately positioned and must protrude 8.5 mm above cylinder head. (*NEA*)

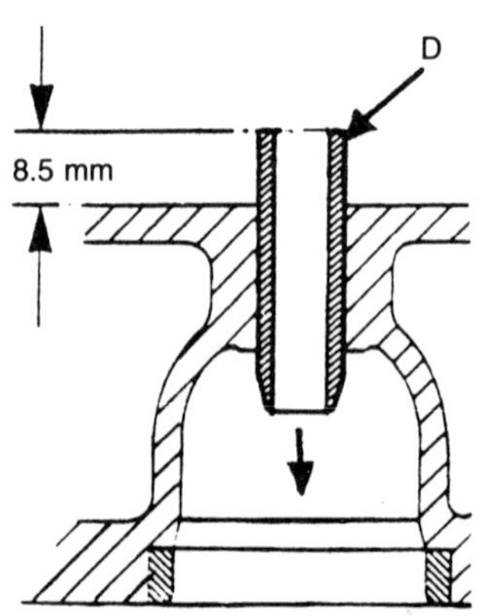

People and places

From the early days to modern times transport has had an effect on people and places. Convenient, efficient and reliable transport has changed work patterns, leisure activities and have had environmental effects.

41. History

To write a complete history of the car would take a long time. Here is a brief outline of the main events.

1769 Nicholas Cugnot made a steam-driven, three, wheeled carriage on a wooden chassis, with the boiler and driving gear on the front wheel. It was used for hauling artillery and ran at 4 mph for 15 minutes.

1801 Richard Trevithick made a steam carriage for 8 people.

1830 More people became interested in the new steam vehicles. The sparks, fires and noise annoyed people so much that the railway and horse-coach owners were able to force steam coaches to be banned. Some engineers continued making steam carriages for a while.

1865 The Red Flag Act virtually stopped the development of cars in Britain. It was an Act of Parliament that said that all vehicles had to have a crew of two and a third person to walk ahead with a red flag to warn other road users. It also imposed a speed limit of 4 mph.

Fig 41(b) Model T Fords (1910)

1860 On the continent engineers continued to develop engines and vehicles. Lenoir used the first internal combustion engine in a vehicle.

1876 Count Nicholas Otto built an engine that used a four-stroke cycle. It was powered by gas and was not used in a vehicle.

1878 Daimler (who had been working for Otto) built a ½ h.p. petrol-driven motor cycle.

1885 Benz was also working on petrol engines and made the first practical four-stroke petrol-engine driven vehicle. It was a three-wheeled vehicle with tiller steering and a top speed of 8 mph which he started to sell and 'cars' became more widely available.

Daimler and Benz continued to improve their engines and Panhard and Lavassor used a Daimler engine in a front-engine rear-wheel drive layout. The French engineers made use of the good roads that had been built in France by Napoleon and began to build cars

Fig 41(a) Daimler's first four wheeled vehicle (1878)

Fig 41(c) BMC Mini (1959)

with large wheels, high seats and petrol engines which gradually began to look a bit less like a horse carriage without the horse.

There were several developments in engineering that made petrol driven vehicles a practicality.

1 Petrol was available as a fuel.
2 Good quality iron and steel was being produced.
3 Machines were developed which could accurately machine cylinders and shafts.
4 Electrical devices such as coils and spark plugs were developed.
5 Ackerman steering and differential gearing were developed.

1896 The Red Flag Act was repealed and the speed limit raised to 12 mph. Motorists held a rally to celebrate and drove from London to Brighton. British engineers began working on cars. Engines became more reliable and were put under a 'bonnet'; windscreens and lights were added and passenger comfort became more important as more people bought the cars that Lanchester, Wolseley, Rover, Riley and Rolls-Royce were building.

1908 Henry Ford decided to make a cheap car that many people would be able to afford. He used an assembly line to 'mass produce' the Model T Ford. He only made one model in one colour but it was cheap, reliable and spares were available (parts were interchangeable and did not have to specially made by individual engineers). The Model T cost £240 and 15 million were sold in 15 years. Cars were becoming cheaper than horses to buy and maintain, steam cars were unpopular and the four-stroke petrol engine powered car became an accepted form of transport.

1920 Morris, Cowley and Austin used mass production techniques to make cheap cars in Britain. Engineers began to think about body styling.

1930 The smaller firms began to go out of business as they could not afford to compete with the large, well equipped production lines making economical and reliable pressed steel-bodied cars. There were two million cars in Britain.

Fig 41(d) An early Vauxhall (1911)

Fig 41(e) An early American tractor

1940 American designers produced bigger and better cars.

1950 Safety became more important. British designers 'copied' American styling. Different types of cars, in different colours with better equipment were produced. Roads were improved.

1960 European (especially Italian) designers produced smaller cars with more rounded body lines.

1970 Designers concentrated on improved styling, safety, comfort and economy.

1980 Japanese manufacturers produced very well equipped cars at very low cost. Fuel economy became very important as well as pollution control and safety.

QUESTIONS

1 List the factors which led to the petrol-driven vehicle being developed.

2 Write a short 'History of the Car'. Mention horses, steam, Red Flag Act, Otto, mass production, design and styling, safety and pollution.

42. A geographical history

Fig 42(a) Pre 1880 early cars were made in Britain, France and Germany

Fig 42(e) 1910–1920 Ford fac

Fig 42(b) 1880–1890 cars were experimental and temperamental

Fig 42(c) Early roads were cart tracks

Fig 42(d) 1900–1910 road racing in France improved performance and reliability

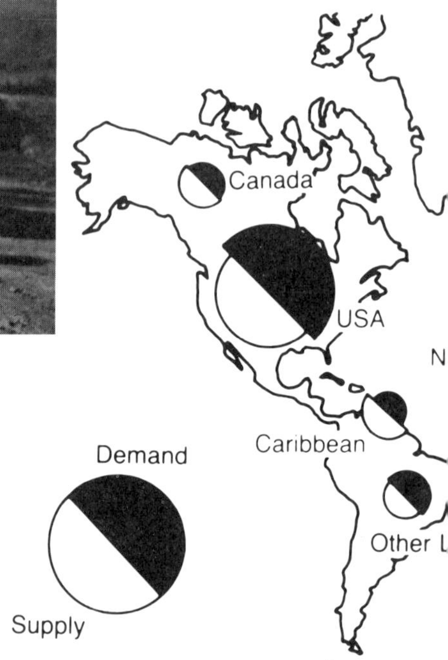

Fig 42(f) C advertisen aimed at families in Britain

ɘ beginning of mass production in America

Fig 42(h) 1940–1950 Military vehicles develop in Europe and America

World oil supply/demand by country

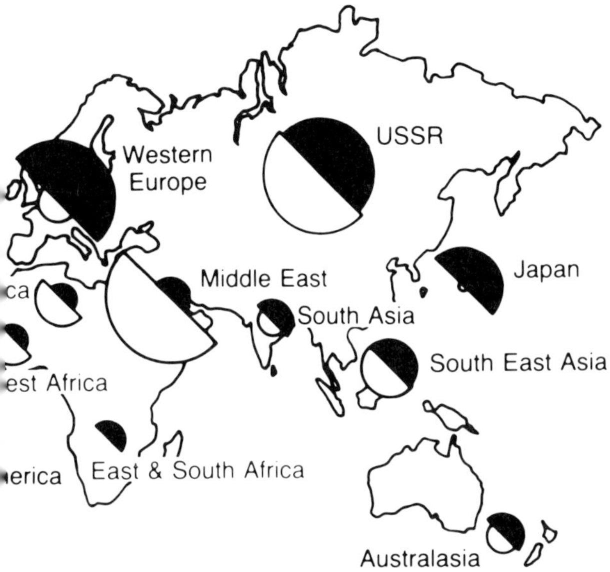

Fig 42(i) 1950–1960 Cadillac – typical of American styling of the time

Fig 42(j) 1960–1970 Italian car designers developed smoother lines and fuel economy

Fig 42(g) 1930–1940 VW Beetle – the Volkswagen (people's car) in Germany

Fig 42(k) 1970 Japanese cars increased competition with low costs, many 'extras' and good fuel economy

QUESTIONS

1 Make a neat sketch map of the world and add notes to explain the part played by various countries in the development of transport.
2 Write a short essay to outline how cars developed in Europe.

43. Motorcycles

Motorcycles provide economical transport. They are relatively inexpensive, simple to maintain and economic to run, and also avoid traffic jams and parking problems. The cheapest are usually single-cylinder two-strokes which offer low fuel consumption at low speeds for short-distance journeys. Long-distance journeys at higher speeds (and higher fuel consumption) require the additional features offered by more expensive machines.

Fig 43(d) Motorcycles and sidecars (1940)

Fig 43(a) Daimler's motorcycle

Fig 43(c) Army despatch rider

Fig 43(e) The British Norton (195

Fig 43(f) Cheap transport for young Italians

Fig 43(g) The Japanese influenc

Fig 43(b) An early American motocycle

Fig 43(h) Multi-plate clutch

A brief history of motorcycles

1880 Daimler put an engine in a motorcycle before he put one in a car.

1920 Engines were put into what were basically bicycle frames.

1940 Motorocycles were put to a variety of uses during the war and after the war they continued to offer cheap family transport and racing that was much cheaper than car racing.

1950 Britain had built a motorcycle industry but failed to develop new models and began to lose money when Italian scooters appeared and became popular.

1960 The Japanese motorcycle industry began to take over with small 50cc commuter motorcycles and, later, bigger, more sophisticated machines.

Technical features

There are several features which distinguish motorcycles from cars in terms of the way that they are engineered. Often they have multiplate clutches and chain drive. The engines may be either two-stroke (cheap, light but with a slightly higher fuel consumption and uneven running at low speeds) or four-stroke. Other differences include suspension, brakes and steering.

Clothing

Because the rider is exposed to the weather and has to be protected against harm in the event of an accident, protective clothing is essential. Crash helmets are a legal requirement; reflective tape is advisable so that others may see the rider at night; leathers reduce the risk of cuts if the rider falls; waterproofs are important in British winters!

Fig. 43(i) Chain drive

Fig 43(j) Shaft drive

'ig 43(k) Disc brake

Fig 43(l) Suspension

.g 43(m) Carburettor

Fig 43(n) Weather proof protective clothing

Fig 43(o) A motorcyclist's leather protective clothing

Fig 43(p) A racing machine

Fig 43(r) A 250cc engine

Fig 43(q) 125cc dirt track bike

Accidents

There is no doubt that the rider of a motorcycle is very vulnerable in an accident. In an 'argument' with a car the motorcyclist is more likely to be killed or seriously injured, whether through his error or that of the other driver. Many motorcyclists are young. 16 to 17 year olds are able to ride machines up to 50 cc; over 17 the limit is 250 cc until the rider has passed the two-part driving test. RAC and other riding schemes help produce better riders but the accident statistics still show a considerably higher risk for motorcyclists than car drivers.

Graph to show the relationship between age/type of vehicle and deaths and injuries

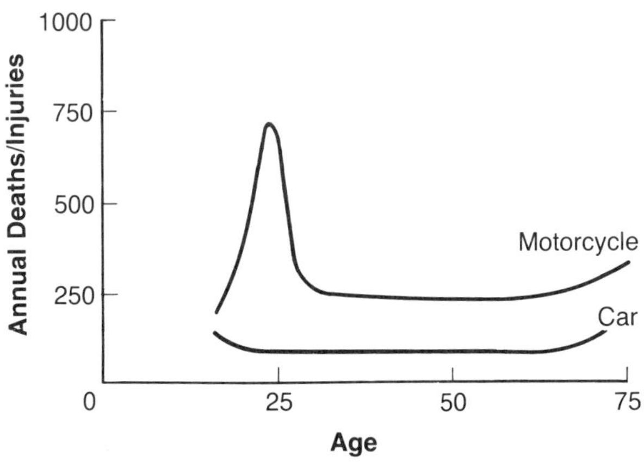

QUESTIONS

1. Write, in note form, a short history of motorcycling.
2. Write a short essay about motorcycle safety.

44. Buying a car

Getting the facts

Before buying a car, find out as much as you can. Talk to friends, family, mechanics, people who own the sort of car that you want. Read magazines, newspapers, road test reports, such as those in AA/RAC books and *Which?* magazine.

New or second-hand?

New cars cost a lot to buy and insure but are guaranteed and reliable. Second-hand will be cheaper but might not be guaranteed and may need repairs or new tyres, battery, exhaust etc. Remember that the person selling a car is selling for a reason and wants your money! Find out why they are selling.

Bear in mind that while expensive cars lose value through depreciation, cheap cars may have faults which are costly to put right. You will get some idea of average prices by reading local papers, magazines and car price guides.

To pay for the car, you might

- Save up and pay cash
- Borrow from a bank and pay interest
- Buy on Hire Purchase (HP) and pay more interest.

TOTAL COSTS INVOLVED

- The price of the car,
- Interest on the loan if you have borrowed from a bank or bought on HP,
- Vehicle licence (tax),
- Insurance (ask for a quotation),
- Any necessary repairs, maintenance, new parts,
- Petrol, oil.

Department of Transport test (the old MOT)

If a car is three years old or more it must be tested by a garage approved by the Department of Transport (previously the *M*inistry *O*f *T*ransport). It is a legal requirement and has to be carried out every year. Brakes, lights, steering and dangerous corrosion are checked. The test keeps unsafe vehicles off the roads but engine, gearbox, battery etc. are not checked and it is *not* a guarantee of the condition of the vehicle.

"If it's a foreign car, what are the costs and availability of spare parts?"

"Second hand cars are cheaper but make sure someone who knows about cars looks at it first."

"Save up and pay cash for your new car."

"Is it the car for the purpose you have in mind."

"Check the car body for signs of rust."

"What is the petrol consumption?"

"If buying second hand, look through magazines and newspapers to get an idea of current prices."

"Is it comfortable to sit in and drive on long journeys?"

"See what discount you can get for paying cash."

"AA/RAC and 'Which Car?' magazines will provide information about new cars."

QUESTIONS

1 List the advantages and disadvantages of buying from: friend, dealer, private advertiser, auction.

2 Make a list of things that should be checked when buying a used vehicle. Use the headings: bodywork, mechanical, electrical, miscellaneous.

45. Documentation

Driving licence

Learner drivers have a provisional licence and must display L-plates. In a car they must be accompanied by a qualified driver. Once the driver has passed a driving test a full licence may be obtained. The application form for a licence will ask for details such as name, age, address, physical health etc. If the driver commits a driving offence the licence may be 'endorsed' and if three endorsements are made then the licence will be withdrawn.

Road Fund Licence

This is the 'car tax' which covers either 6 months or one year. The 'tax disc', which is issued upon payment of the tax, must be displayed on the car windscreen.

Insurance

This is necessary to pay for damage or compensation for injuries received in any accidents that may occur. There are three 'parties' in most insurance claims: the driver, the insurance company and the 'third party' (the other driver). The legal minimum insurance that all drivers *must* have is 'Third Party' insurance so that any person who wants to claim from the driver may make a claim. The three main types of insurance are:

1 *Third Party only.* The legal minimum, which covers only claims by a third party against the driver.
2 *Third Party, Fire and Theft.* This covers third party claims and insurance against the vehicle catching fire or being stolen.
3 *Comprehensive.* This covers almost everything that is likely to happen: third party claims, fire, theft, accidents to the driver's vehicle in which no other vehicle is involved, etc.

NO CLAIMS BONUS

A 'premium' is paid each year for the insurance. If the driver has no accidents that result in a claim against his insurance company then his premium for the following year will be less. This is called a 'no claims bonus'. If a claim is made then the cost of insurance will go up as the driver loses this 'bonus'.

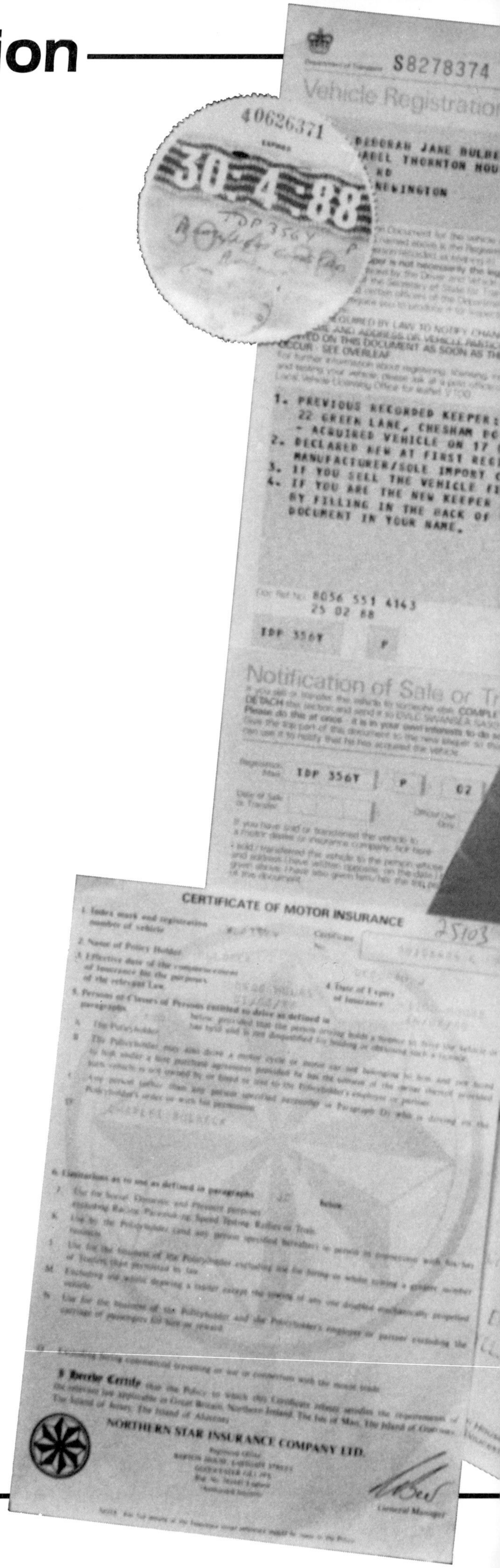

Registration document

This states who is the authorised owner of the vehicle. It has replaced the old 'log book'.

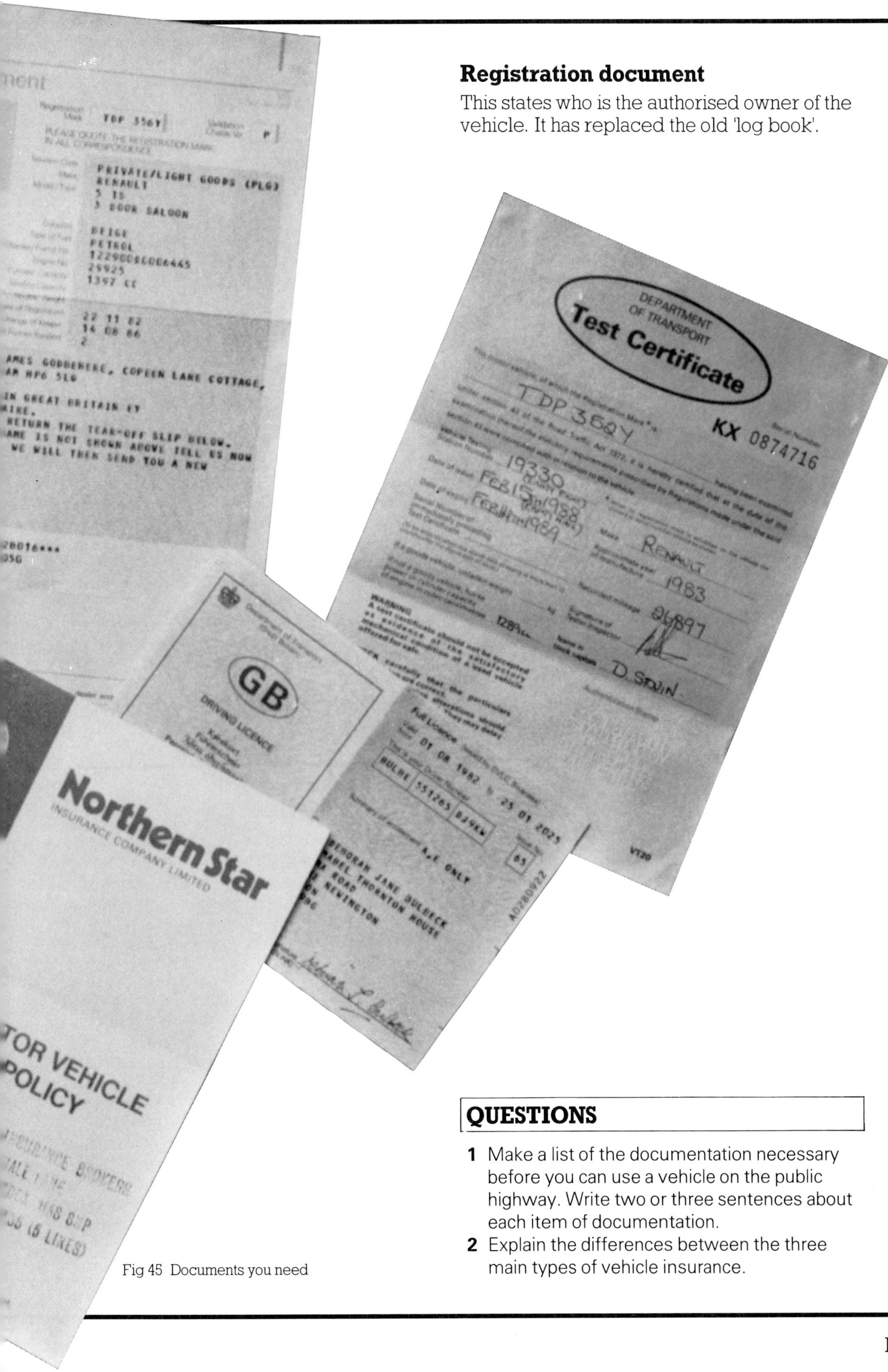

Fig 45 Documents you need

QUESTIONS

1. Make a list of the documentation necessary before you can use a vehicle on the public highway. Write two or three sentences about each item of documentation.
2. Explain the differences between the three main types of vehicle insurance.

46. Pollution and exhausts

Pollution from exhaust gases

Exhaust gases can harm people, plants and the environment. In America and Europe there are regulations to reduce the amount of pollution caused by exhaust emissions. American cities such as San Francisco have suffered from exhaust gases causing thick fog (called smog) which is dirtier than ordinary fog and more harmful to health.

HARMFUL EXHAUST EMISSIONS

- *Carbon monoxide* from partially burnt fuel is poisonous.
- *Hydrocarbons* from unburnt fuel are irritants which may cause cancer.
- *Carbon* from partly burnt fuel causes smelly smoke which discolours buildings.
- *Nitrogen oxides* formed during combustion contribute to breathing problems and to smog.
- *Lead*, added to petrol to improve its octane rating, causes blood and brain disorders in humans.

WHAT CAN BE DONE?

Devices can be fitted to reduce the pollution but they are costly and often reduce power. Three main ideas are:

1 *Better engine tuning.* This means that more of the mixture is burned and so less carbon monoxide, hydrocarbon and carbon enters the atmosphere.

2 *Lead-free petrol.* Taking the lead out of petrol will keep the air free of a very dangerous pollutant.

3 *Devices such as catalytic converters.* These convert unburnt fuel (hydrocarbon) and carbon monoxide into carbon dioxide and water which are harmless. They do this by a process called oxidation.

'HC' + CO + O_2 combine to form $CO_2 + H_2O$.

Exhaust systems

Exhaust gases contain carbon dioxide (which can suffocate) and carbon monoxide (which is poisonous) so they must not enter the car. The exhaust system is the manifolds, pipes and silencers which take exhaust gases from the cylinder to the atmosphere.

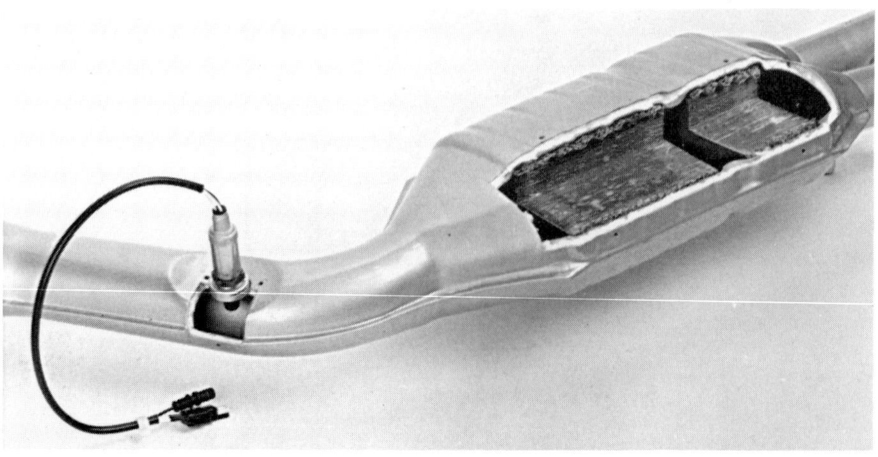

Fig 46 Catalytic converter

MANIFOLD
The manifold collects gases from each cylinder into usually one pipe. Sometimes it is a one-piece cast iron manifold which bolts to the side of the cylinder head and finishes in a circular flange to which the exhaust pipe is fitted. Branched manifolds give a better gas flow and cause less 'backpressure' but they cost more.

SILENCER
Sounds are air movements. The rapid burning in the cylinder causes very rapid expansion and movement of the gases and this creates a lot of noise (which is a form of pollution). To reduce the noise the air must be slowed down. Metal plates called baffles slow the airflow but caused a backpressure which reduces engine power slightly. A large expansion chamber allows the gases to expand and slow down and creates less backpressure. A straight through silencer uses glassfibre to absorb the sound.

Baffle silencer: cheap but cause backpressure and reduce performance. Expansion or resonance chamber silencer: less pressure but more expensive. Straight through silencer: least effect on performance but not very quiet!

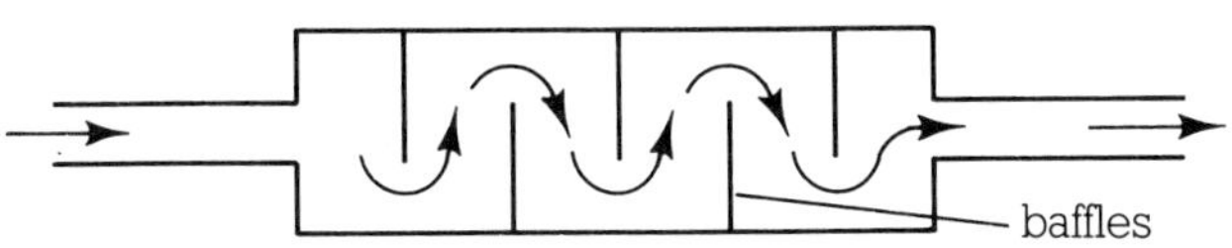

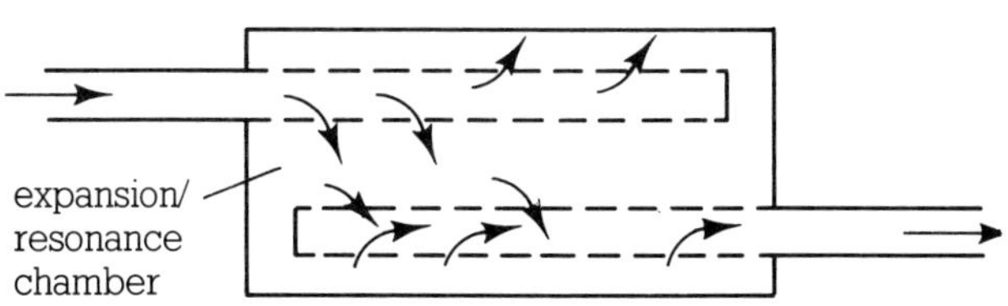

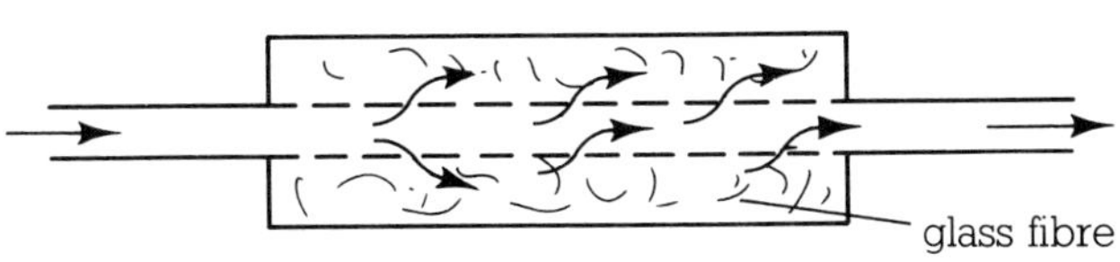

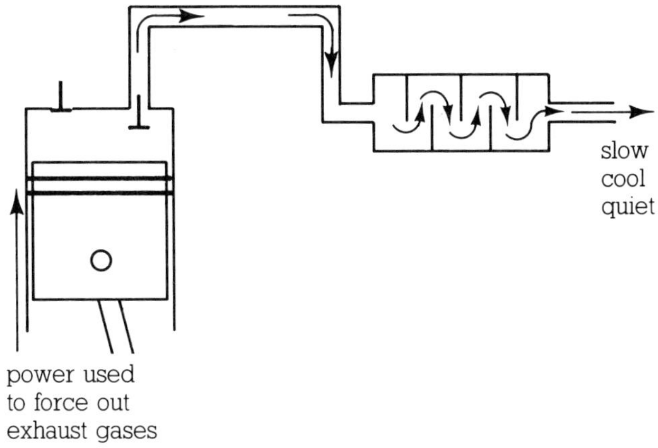

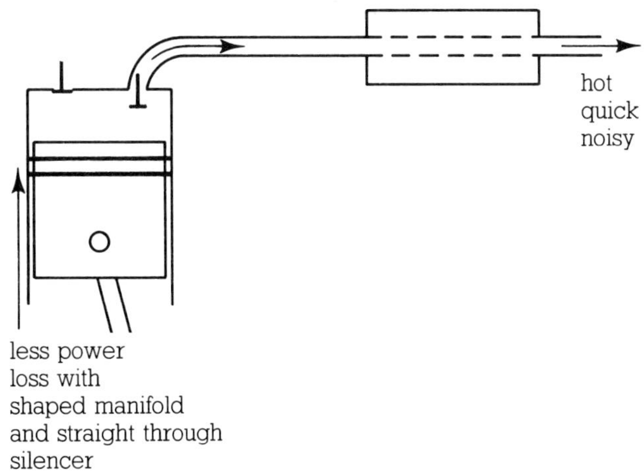

RUST
Exhaust systems rust from the inside and the outside. Mild steel rusts very quickly on the outside in rain, salt, and mud. (Salt is used on roads in winter to clear the ice and is then thrown up onto the bottom of the car.) The inside of the system is corroded by combustion products. When petrol burns it forms lead salts, water vapour and acids. When the system cools these condense on the inside of the exhaust and the acids attack the steel.

Mild steel is used because it is cheap; stainless steel exhaust systems last a lot longer but cost more.

QUESTIONS

1. List the main pollutants present in exhaust gases and explain what can be done to reduce pollution.
2. Make neat labelled diagrams to show the three types of silencer.

47. Energy conversions and conse

Internal combustion engines work by changing chemical energy (stored in the fuel) into heat energy (when it burns) and then into mechanical energy (as the expanding gases push the piston down). There are other energy conversions which occur in various parts of the vehicle.

Petrol is obtained from oil, which is a fossil fuel that has taken thousands of years to form. Once the reserves of oil have been used we shall not have petrol as a fuel for car engines. (We shall not have plastics either, as these are also oil products.) For this reason people are concerned about energy conservation – to make better use of the fuel which we have so that it may last longer.

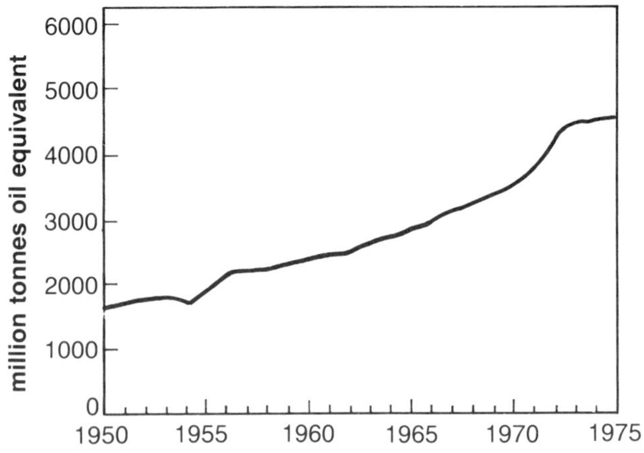

increase in oil demand 1950-1975

Fuel conservation

Petrol consumption increases with speed. You can save petrol by driving 10 mph slower on a journey without adding very much to your journey time. Overdrive units reduce engine speed compared with road speed and can pay for themselves in a short while by reducing fuel bills. Faulty spark plugs, poor timing, carburettion problems etc. all waste fuel and money and are cheap and easy to correct.

Low viscosity oils create less drag and therefore reduce fuel consumption. Diesel engines use only 60%–70% of the fuel that petrol engines use, and are widely used by high mileage vehicles such as taxis and commercial vehicles. Additional simple methods of reducing fuel consumption are checking tyre pressures and wheel alignment, driving with an empty boot and the windows closed, removing roof racks when they are not being used, and so on.

Energy conversions

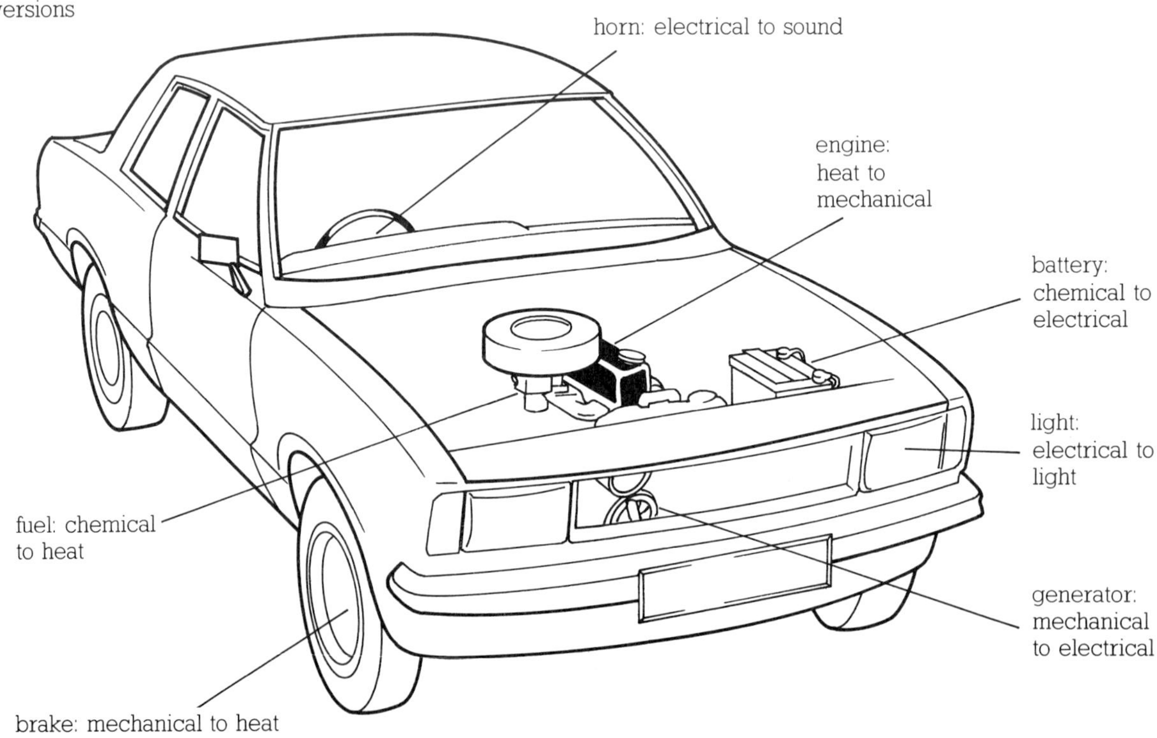

ation

DRAG COEFFICIENT

The shape of the car affects the way in which air flows over it and this affects performance and economy. The resistance to airflow is called 'drag' and is measured as a drag coefficient. Early cars were very high and flat fronted. This gave them a very high drag coefficient but the speeds were low and drag was less important. The low, smooth, streamlined shape of modern cars means lower drag coefficient and lower fuel consumption.

Fig 47(a) Low drag coefficient

Fig 47(b) High drag coefficient

Rusting

Cars are made from steel, which is made from iron. Iron is formed by the reduction of iron oxide ore. Rusting is a reversal of this. Iron and oxygen combine to form iron oxide and the reddish-brown powdery substance formed by this oxidation is called rust. Rust is another example of a valuable resource being lost.

Fe + O combine to form FeO

Iron rusts when it is in contact with oxygen and water. It can be prevented by:

- Painting, which excludes moisture (until it is scratched!).
- Galvanising, which means coating it with a non-rusting metal such as zinc.
- Plating with a metal such as chromium, which tends to be expensive.
- Undersealing, which means using a thick bitumen based 'paint'.

Rusting

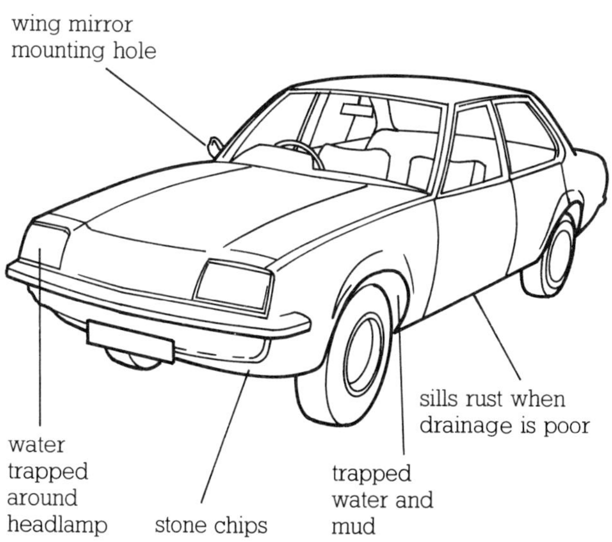

QUESTIONS

1 Write a short summary of measures that may be taken to conserve fuel.

2 Explain what is meant by rusting, how it takes place and what may be done to prevent it.

48. Social effects

Road transport has employed many people in building vehicles and roads, maintenance and servicing, sales and support, service industries, driving, deliveries etc. It has had the benefits of providing more people with the opportunity to become mobile – they are able to drive into the countryside at weekends and to travel to shops and offices. Out-of-town housing has developed with 'commuter dormitory towns' where people sleep and spend their weekends; while for the remainder of the time they are at work many miles away.

The social benefits of increased car ownership has to be balanced against the disadvantages of the intrusion of people into quiet country areas, increased noise and atmospheric pollution, damage to the countryside in road building programmes and the high number of deaths and injuries which occur every year.

Fig 48(c) Rural villages sometimes suffer

Fig 48(a) A social benefit

Fig 48(b) Supermarkets serve people from a wide area

Fig 48(d) Demonstration against a new motorway

Fig 48(e) An America motorway scene

Motorways

Motorways have developed in Britain since 1960 (later than in Germany and America) into a network which allows swift transport of goods and people. They are more expensive to build than conventional roads, use a great deal more space and result in pollution which conservationists have objected to. However, they offer a good distribution network for goods and the industrial sites in South Wales rely heavily on the M4 and the Severn Bridge. An area such as Bristol, at the intersection of two motorways, makes an ideal site for factories and warehouses and this brings more jobs to the area.

Another positive feature of motorways is that they have made leisure driving quicker, easier and safer. Traffic has been taken away from small villages and congested town centres. People are carried quickly over long distances and holiday areas such as Cornwall and Scotland have come within the reach of more people.

Commuter traffic using motorways is able to get to and from city centres far more easily than along traditional roads and although the traffic is heavy it usually is kept moving.

Freight traffic on the roads has some opponents and large lorries do cause holdups and increased wear and tear on the road surface but the low cost and convenience has meant that road transport often competes very well against the rail network that was built up in Victorian times by people such as Brunel.

Fig 48(f) A motorway intersection

Fig 48(g) Severn bridge

Fig 48(h) Motorway intersections use large areas of land

Fig 48(i) Commuter traffic speeding out of a city centre

QUESTIONS

1. Write a short essay on the social effects which widespread car ownership has had in this country.
2. List the advantages and disadvantages of goods transport by road and goods transport by rail.

EXAMINATION QUESTIONS

PEOPLE AND PLACES

1 What was the Red Flag Act?

2 Why are motorcycles often a popular alternative to cars as a means of transport in towns?

3 Why is the Model T Ford an important car in the history of motor vehicles?

4 What are the main items of protective clothing that a motorcyclist needs?

5 Which one of the following statements is true?
   a) In the 1890s four cylinder car engines were common.
   b) Pneumatic tyres were invented and fitted to some horse-drawn carriages by John Boyd Dunlop in 1945.
   c) The earliest self-propelled road vehicles capable of carrying people employed steam engines.
   d) The German pioneer Doktor N. A. Otto first developed the idea of compressing the charge of gas and air within the internal combustion engine prior to its ignition. (*LEA*)

6 What are the main effects that the motor vehicles have as sources of pollution to people, buildings and the general environment?

7 Motor vehicle ownership has had many social effects. Describe the positive and negative effects and explain your own views on the future of 'motoring for the masses'.

8 When applying at a Post Office for a new road tax disc for a five-year-old car which one of the following documents would *not* be needed?
   a) Certificate of motor insurance;
   b) MOT certificate;
   c) Driving licence;
   d) Application form. (*LEA*)

9 The illustration shows three alternative routes for a proposed motorway. Discuss the advantages and disadvantages of each in terms of cost, technical problems, environmental effects, effects on industry and the local population and other points that you feel are important.

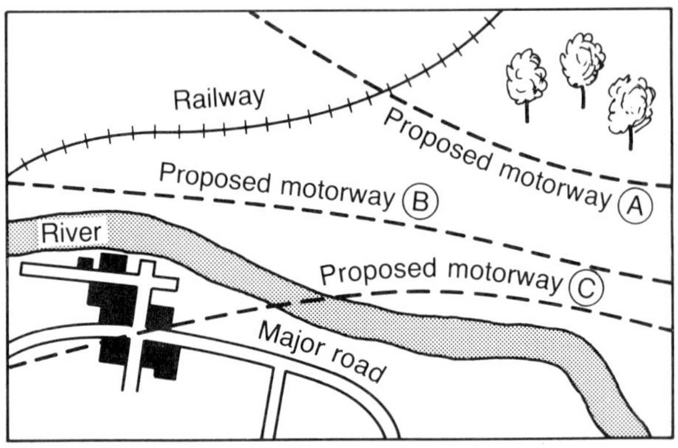

10 Describe the part played by each of the following countries in the development of motor vehicles:
   a) Great Britain
   b) France
   c) Germany
   d) Italy
   e) U.S.A.

Abbreviations

ac	alternating current
bdc	bottom dead centre
bhp	brake horse power
BSF	British Standard Fine
BSW	British Standard Whitworth
CAD	Computer Aided Design
CAM	Computer Aided Manufacture
cb	contact breaker
ci	compression ignition
dc	direct current
DoE	Department of Environment
DoT	Department of Transport
fdu	final drive unit
HT	High tension
LT	Low tension
MoT	Ministry of Transport
mpg	miles per gallon
mph	miles per hour
ohc	over-head camshaft
ohv	over-head valve
rpm	revolutions per minute
SAE	Society of Automotive Engineers
tdc	top dead centre
uj	universal joint

Index and Glossary

four-stroke cycle 8
sequence of induction, compression, power, exhaust used in most car engines

fuel filter 18
removes impurities from fuel before it passes into the carburettor

fuel tank 18
pressed steel container which stores fuel

fuse 72
a thin wire which melts if too high a current flows through a circuit

gear box 56
contains gears which allow speed of road wheels to vary compared with engine speed

generator 27
converts mechanical energy into electrical energy (see alternator and dynamo)

glow plug 14
used in compression ignition engines for starting from cold

grease 47
lubricating oil mixed with soap, used in bearings etc.

handbrake 63
mechanical brake used for parking

Hotchkiss drive 66
propeller shaft connected to rigid rear axle mounted on leaf springs

hydraulic system 62
operation of brake or clutch using fluid flowing from master cylinder to slave cylinder

hydrometer 76
instrument used to measure the relative density of electrolyte in a battery

ignition 26
burning of fuel in an engine

injector 14
injects diesel fuel into compression ignition engine

insulator 73
material which prevents the passage of electricity

internal combustion 6
burning of fuel/air inside a cylinder

leaf spring 66
spring made of several layers of spring steel

lubrication 48
use of oil or grease to reduce friction

McPherson strut 67
simple and effective front suspension system

mechanical fuel pump 19
driven by gear on camshaft, it pumps fuel from the fuel tank to the carburettor

octane 25
the resistance to detonation in petrol

oil 46
liquid lubricant used in car engines

oil filter 49
removes impurities from lubricating oil

Otto cycle 8
see four-stroke engine

petrol 24
fuel which vaporises easily and burns evenly

petroil 49
mixture of petrol and oil used in two stroke engines

piston 10
aluminium alloy component which slides inside cylinder

piston ring 10
makes gas-tight seal between piston and cylinder wall

pneumatic tyre 64
tyre which is filled with air

port 12
passage from combustion chamber to inlet or exhaust manifold

primary safety 92
design features which reduce the risk of an accident occurring

propeller shaft 58
connects gearbox to final drive unit

rack and pinion 68
steering system in which a pinion gear meshes with a rack gear

radial-ply tyres 64
tyre in which cords are at right angles to the casing

radiation 43
heat travel through a vacuum

radiator 44
stores and cools water

rotary engine 15
uses a triangular rotor in a kidney shaped cylinder

rotor arm 35
rotates inside distributor passing H.T. current from coil to each spark plug in turn

rusting 115
iron combining with oxygen to oxidise and form iron oxide

sealed beam unit 74
headlight with filament, reflector and lens in one sealed unit

secondary safety 92
design features which reduce effect of an accident on driver and passengers

shock absorber 67
reduces effect of uneven road surface on steering and comfort

sparking plug 30
converts high tension voltage into a spark to ignite the fuel/air mixture

starter motor 74
turn flywheel and crankshaft when starting engine

sump 48
pressed steel compartment for storing oil in wet sump lubricating system

thermostat 45
prevents cooling water from circulating to radiator until engine has warmed up

transfer port 13
passageway through two-stroke engine which connects crankcase to combustion chamber

transmission 58
the system which passes torque from the engine to the wheels

two-stroke engine 12
uses both sides of the piston to complete a cycle in only two strokes of the piston

viscosity 46
the 'thickness' or resistance to flow of a liquid

Wankel engine 15
see rotary engine

water pump 44
driven by fan belt, it pumps cooling water around an engine

wet sump 48
lubrication system in which oil is stored in a sump and pumped to components